Safety and Health Considerations for the Design of Fire and Emergency Medical Services Stations

FEDERAL EMERGENCY MANAGEMENT AGENCY
UNITED STATES FIRE ADMINISTRATION

SAFETY AND HEALTH CONSIDERATIONS FOR THE DESIGN OF FIRE AND EMERGENCY MEDICAL SERVICES STATIONS

This publication was produced under contract EMW-95-C-4703 for the United States Fire Administration, Federal Emergency Management Agency. Any information, findings, conclusions, or recommendations expressed in this publication do not necessarily reflect the views of the Federal Emergency Management Agency or the United States Fire Administration.

PREFACE

This manual was developed under contract for the U.S. Fire Administration to provide comprehensive guidelines for the design or remodeling of fire and emergency medical services (EMS) stations and other facilities (e.g., training centers) in terms of safety and health concerns. The purposes of this manual are:

1. To alert fire and emergency medical service personnel to potential safety and health hazards within the station and other facilities;

2. To identify pertinent regulations which affect the construction and inspection of fire and EMS stations which can be applied to station design for safety and health;

3. To establish compliance guidelines for new station construction/existing station modification with model specifications that can be adopted as part of a fire/EMS department's station design/construction bid package; and

4. To provide a checklist for station health and safety inspections and to assist evaluations of existing station designs.

This is not a manual which provides a detailed step-by-step procedure for designing fire or EMS stations. Rather, this manual is intended to be a useful guide when addressing design aspects of the fire or EMS station which pertain to the health and safety of personnel who must work and live in those facilities.

It is recognized that many of the nation's fire houses and EMS stations were built prior to the development of this document. Therefore, this manual should be consulted for guidance both when facilities are remodeled and new ones are built. Often city managers, emergency medical directors, fire chiefs, and local government officials are faced with a variety of choices when undertaking new construction or remodeling. This manual has been prepared to simplify this process by providing a comprehensive list of requirements and suggested design alternatives.

Users of this manual are encouraged to implement, manage, and develop safety and health concepts for the well being of their emergency service personnel when designing and building fire and EMS facilities, The guidelines in this manual are also intended to help reduce actual and potential accidents and injuries in the fire or EMS station.

The information provided in this manual was developed from a variety of references including consensus standards, OSHA regulations, occupational health codes, provisions of the American Disabilities Act (for public areas) and related articles relative to the construction of such facilities. Recommendations in this manual are primarily based on Federal OSHA regulations. Users of this manual are encouraged to determine the applicability of station standards, local codes and ordinances, consensus standards, and recommended practices for their particular area.

ACKNOWLEDGEMENTS

The authors of this manual would like to acknowledge and thank several individuals who were involved in this project and who provided valuable input for assisting in the development of the this manual.

For this project, a Quality Review Panel, consisting of experts with different backgrounds to provide broad perspectives, served as the primary resource in the manual's development. These individuals included:

- Sherri-Lynne Almeida, Houston Fire Department, Houston, Texas
- John Ball, John Ball and Partners Architect, Inc., Phoenix, Arizona
- W. Robert Barnard, Washington State Office of Fire Protection Services, Olympia, Washington (retired)
- Richard M. Duffy, International Association of Fire Fighters (IAFF), Washington, DC
- Jim Minx, Oklahoma State Firefighters Association, Oklahoma City, Oklahoma
- Thomas Thorpe, Sandy Springs Volunteer Fire Department, Maryland; Litton Systems, Inc., College Park, Maryland

The International Association of Fire Fighters (IAFF), Fire Department Safety Officers Association (FDSOA), National Fire Protection Association (NFPA), and National Volunteer Fire Council (NVFC), the Volunteer Firemen's Insurance Services, Inc. were helpful in providing useful information.

In addition, several organizations responded to requests for specific information which were helpful to the project, including:

- Austin Fire Department (Texas),
- City of Camden Fire Department (New Jersey),
- City of Glendale Fire Department (California),
- Dallas Fire Department,
- Fairfax Country Fire and Rescue Services (Virginia),
- Los Angeles City Fire Department,
- Marion Fire Department (Iowa),
- Tulsa Fire Department, and
- U.S. Coast Guard Facilities Design and Construction Center (Pacific).

TABLE OF CONTENTS

SECTION 1 - INTRODUCTION

U.S. Fire Administration injury statistics show a significant number of injuries, particularly strains and sprains, have occurred away from the emergency activity. [1] Many of these injuries and deaths are occurring where they would be least expected. Fire or emergency medical service department facilities are rarely looked at in terms of their *potential* safety and health hazards. Yet according to U.S. Fire Administration fatality summaries from 1983 to 1995 prepared by the National Fire Protection Association (NFPA), a total of 17 firefighters have died at fire stations from causes other than cardiovascular systems deaths. [2] These deaths include:

- seven falls (including two from hose towers),
- three carbon monoxide poisonings,
- two crushing traumas due to vehicular movement,
- one electrocution,
- one steam boiler explosion,
- one SCBA cylinder explosion,
- an overturned tractor crushing trauma, and
- one homicide,

At least two of the firefighters died in a hose tower accident while hanging wet hose. Hose tower ladders are known as a common fall hazard that can result in a traumatic accident in the fire station (yet there are safety regulations which address proper fall protection requirements). Other common areas with a high risk of injury in the fire/EMS station are sliding poles, apparatus bays, and battery charging rooms.

Firefighter and emergency medical service (EMS) personnel injuries at the station are much more common than recognized by the industry. [3] Unfortunately, firefighter injury statistics specifically for station injuries are not collected consistently on a national level. [4] However, this information can be obtained for individual departments when injury information is maintained in computer databases that allowed searching by location. Figure 1 shows station injury data by injury type and affected body part for a large and moderately sized fire department. Table 1 shows the causes for station injuries over a three year period. The costs associated with these injuries are enormous. One large fire department has estimated that its spends over $1 million annually for injuries occuring in its stations. Clearly these data, though limited, point to the seriousness of, and need for, improved station design from a safety and health perspective.

---

[1] Statistics from the National Fire Incident Report System (NFIRS), U.S. Fire Administration Fire Coordination and Data Analysis Branch, Emmitsburg, Maryland.

[2] Firefighter Fatality Reports prepared by the National Fire Protection Association for the U.S. Fire Administration, 1983 to 1995.

[3] *NFPA* Journal, November/December issues, 1991 - 1995.

[4] Typically station injuries are included in the "other on-duty" category when departments reported to NFPA or other entities. Station injuries might also fall under the "training" or "responding/ returning" heading, if the firefighter was at a station or on station property.

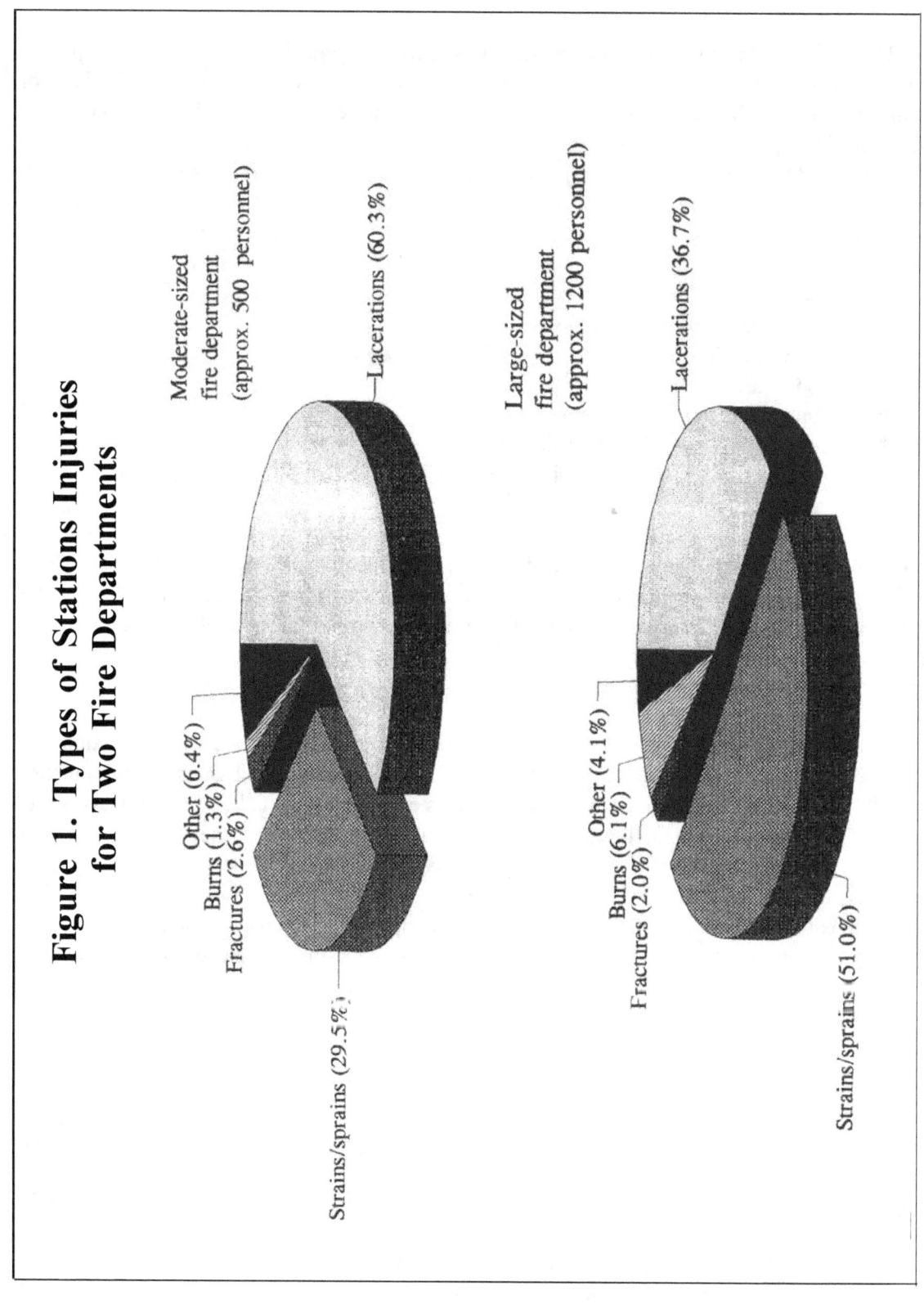

Figure 1. Types of Stations Injuries for Two Fire Departments

Moderate-sized fire department (approx. 500 personnel)

Lacerations (60.3%)

Other (6.4%)
Burns (1.3%)
Fractures (2.6%)

Strains/sprains (29.5%)

Large-sized fire department (approx. 1200 personnel)

Lacerations (36.7%)

Other (4.1%)
Burns (6.1%)
Fractures (2.0%)

Strains/sprains (51.0%)

Table 1. Summary of Station Injury Causes for a Moderate Sized Fire Department, 1992-1994 †

Activity	1992	1993	1994	Total	Rank
vehicle maintenance	42	37	30	109	1
physical fitness activity	40	30	24	94	2
moving about station, normal activity	23	29	19	71	3
moving about station, alarm sounding	23	18	25	66	4
station maintenance	19	12	17	48	5
boarding apparatus	15	11	12	38	6
cooking and food preparation	6	10	7	23	7
sleeping (getting out of quarters)	8	1	8	17	8
equipment maintenance	1	2	11	14	9
other activity	0	3	6	9	10
administrative work	2	4	2	8	11
exiting apparatus	not listed	not listed	3	3	12
physical fitness testing	3	0	0	3	12
showering/personal hygiene	0	1	2	3	12
training activity or drill	1	0	0	1	13
TOTAL	183	157	166	506	---

† Data compiled by a Southwestern fire department with approximately 600 emergency response personnel.

This statistics show vehicle maintenance as the number one leading cause of station injuries, followed by physical fitness (2), moving about the station during normal activity (3) and moving about the station during alarm (4). The average cost to the department for each injury is estimated at $11,200 (includes all associated costs for lost time, medical fees, and workmans' compensation).

Limited injury data point out the types of injuries. Strains, sprains, and lacerations comprise the majority of station injuries. While these injuries are not severe, many cause lost time, and some such as lower back strains, can cause result in extensive lost time and can be debilitating. One moderate sized fire department has determined that each strain or sprain injury of moderate severity costs the department over $25,000 in lost time, workmans' compensation, medical charges, and overtime hiring.[5] Moreover, many types of injuries may go unnoticed. For example, continued exposure to diesel exhaust may partially explain the high incidence of certain cancers among firefighters. [6,7] Although OSHA has confirmed diesel fuel as a carcinogen, these illnesses may be difficult to link to specific exposures at the station, but station environments may be the most likely cause over the long term.

The U.S. Fire Administration has recognized that many emergency response personnel have died and been injured needlessly as a result of accidents at fire stations. This concern is further fostered by the alarming number of occupational health exposures firefighters and EMS personnel have come in contact with in recent years which have been brought back to the station[8,9,10] Federal and state regulations have endeavored to curb exposures to diesel emissions, noise abatement, indoor air quality, hazardous material and waste exposure, plus infectious material, all of which emergency response personnel come in contact with on a daily basis. However, these compliance with these regulations may not be consistent throughout the fire and emergency medical services.

General Requirements

Many emergency response organizations"[11] are familiar with standards of the National Fire Protection Association (NFPA). NFPA 1500, *Standard on Fire Service Occupational Safety and Health Program,* defines a fire department facility as any building or area owned, operated, occupied, or used by a fire department on a routine basis which may include fire and rescue stations, training academies, and communication centers. Fire department facilities do not include those facilities not normally under fire department control. Chapter 7 of NFPA 1500

---

[5]Private communication with Tulsa Fire Department.

[6]Froines John R., William C. Hinds, Richard M. Duffy, Edward J. Lafuente, and Wen-then V. Liu, "Exposure of Firefighters to Diesel Emissions in Fire Stations," *American Industrial Hygiene Association Journal,* Vol. 48, March 1987, pp. 202-207.

[7]Stull, Jeffrey O, , "Controlling Diesel Exhaust Emissions at the Fire Station," *Fire Engineering,* Vol. 147, October 1994, pp. 18+.

[8]Heyer, N., N.S. Weiss, P. Demers, and L. Rosenstock, "Cohort Mortality Study of Seattle Fire Fighters, 1945-1983," *American Journal of Industrial Medicine,* Vol. 17, 1990, pp. 493-504.

[9]Musk, A. William, John M. Peters, and David H. Wegman, "Lung Function in Fire Fighters, I: A Three Year Follow-Up of Active Subjects," *American Journal of Public Health,* Vol. 67(7), 1977, pp. 86-89.

[10]Lewis, S.S., H.R. Bierman, and M.R. Faith, "Cancer Mortality Among Los Angeles City Fire Fighters," L.A. Fire Department Report, Dec., 1982.

[11]This manual applies to station design for the fire and emergency medical services. Fire and emergency medical services are sometimes referred to as emergency response organizations or departments. Likewise, firefighters and EMS personnel may be referred to as emergency response personnel.

on "Facility Safety" requires that department facilities:

- Comply with all legally applicable health, safety, building, and fire code requirements.

- Provide facilities for disinfection, cleaning, and storage in accordance with NFPA 1581, *Standard on Fire Department Infection Control Program.* (NFPA 1581 provides guidelines the recommend against the cleaning and disinfecting of protective clothing and equipment, portable equipment, and other clothing in areas used for food preparation, the cleaning of food and cooking utensils, personal hygiene, or sleeping and living. Also required for disinfection are two sinks with a sprayer attachment, a rack with a drain to the sewer, medical-type non-grasp controls on faucets, and hot and cold water.)

- Provide smoke detectors in work, sleeping, and general storage areas.

- Comply with NFPA 101, *Life Safety Code* or locally adopted requirements of the building code.

- Be designed with provisions for the ventilation of vehicle exhaust emissions from fire apparatus (and other vehicles) to prevent exposure to firefighters and contamination of living and sleeping areas.

- Have designated smoke-free areas including work, sleeping, kitchen, and eating areas.

- Be inspected annually to determine compliance with all legally applicable health, safety, building, and fire code requirements, and that these inspections be documented and recorded.

- Be inspected monthly to identify and correct/document any safety or health hazards.

- Have an established system to maintain facilities and to promptly correct any safety or health hazards or code violations.

- In addition, the U. S. Fire Administration strongly recommends that stations be protected with automatic sprinkler systems.

Emergency response organizations find it difficult to fully comply with the inspection and maintenance requirements because of the potentially large number of requirements which can apply to fire/EMS facilities, the lack of fire/EMS service knowledge concerning station safety hazards, the perception that the station is generally a safer place than the fire ground or emergency scene, and most importantly, the cost of compliance.

There are a significant number of regulations which apply to fire and EMS department facilities. These regulations may be federal, state, or local. Standards such as those from the

National Fire Protection Association may not be mandatory depending on the state or jurisdiction where the facility is located.[12] Then, there are several other sources of non-mandatory regulations. It is impossible to list all the potentially applicable regulations. To do so would create an extensively thick and unreadable reference. For this reason, this manual focuses on mandatory, federal regulations (primarily OSHA) which apply to many states within the U.S. and references other standards and regulations as appropriate. **Federal OSHA applies to those states which do not have an OSHA-approved state occupational safety and health program. States that do have OSHA-approved plans are still required to meet or exceed Federal OSHA standards.**[13] Three appendices containing requirements or listing sources for standards and regulations are provided:

Appendix A   OSHA regulations pertaining to fire/EMS station safety and health
Appendix B   State safety and health standards
Appendix C   List of organizations with standards and information related to station construction

Using This Manual

It is possible to construct a building without identifying what is required by every regulatory agency. Unfortunately when this happens and problems occur, the structure must be modified after it is occupied and functioning. Not only does this often result in unsafe conditions, but the costs for modifying the station are usually greater than if the regulations were considered during the planning of the building. This manual attempts to limit these occurrences by allowing departments to:

1.   identify applicable requirements,
2.   select appropriate design features, and
3.   evaluate their compliance for safety and health.

The primary content of this manual is organized into four primary sections:

Section 2 addresses the planning process. Included in this section are roles and responsibilities of both the department and its building committee as well as the design team. This section focuses on the needs assessment model to determine department requirements and to identify potential hazards and safety concerns in selecting the site and designing the station.

Section 3 provides an overview of general design considerations. Many of the factors described are part of the overall design and construction process; however, specific comments are offered relative to safety and health concerns. Therefore, the principal purpose of this section is to increase safety and health awareness in the selection of specific station designs and features.

---

[12]NFPA 1500 has been adopted by occupational safety and health department in several states, including Florida, Kansas, Louisiana, Maine, Missouri, Ohio, Rhode Island, and Texas.
[13]See Appendix B for a listing of states with OSHA-approved state plans.

Section 4 identifies specific safety and health concerns at the station. Subsections are provided for each safety and health concern that:

discuss the nature of the hazard,
examine the extent of the hazard and its potential severity at the station,
list relevant standards and sources of information, and
provides preventative design requirements for preventing or reducing the hazard.

An initial portion of this section includes a matrix showing station locations where specific hazards are likely. The section ends with a composite table showing specific standards as they apply to the different areas of the station.

Section 5 establishes procedures for conducting a safety and health inspection of new or existing station construction. A comprehensive example checklist is provided in Appendix E which enables the designer or department personnel to review station design for compliance to applicable safety and health requirements.

Basis for Manual

Information presented in this manual was researched through on-site visits to several stations, ranging from rural to metropolitan, to multi-purpose facilities.

Recommendations included in this publication are specifically directed toward those accident/and health concerns associated with activities that occur on fire station premises and that can be reduced by careful design of new and remodeled facilities. Care has been taken to make the suggestions compatible with accessibility regulations and with existing life safety and health requirements.

When appropriate, administrative and procedural considerations or recommendations have been suggested. The architect alone cannot design a safe fire station. It takes the involvement and combined effort of the administration and emergency response personnel to build a station.

SECTION 2 - THE NEEDS ASSESSMENT-BASED PLANNING PROCESS

Adequate planning is the key to a successful station design which addresses all existing and foreseen safety and health hazards. Station planning is an essential step of the design process and should include a number of individuals who are both inside and outside the department. A fundamental part of this process is conducting a needs and risk assessment. The needs assessment involves identifying specific needs for the construction of the station. When needs are adequately identified, then the group responsible for station planning can research and recommend design solutions which satisfy those needs in terms of economy and in meeting safety and health requirements. This section describes a general station design process with a specific focus on planning and provides recommendations for carrying out a general needs assessment.

The Design and Construction Process

The overall design and construction of a fire or EMS station involves several steps and can take as much as three years to complete. Table 2 shows an example schedule for the station design and construction process.

Table 2. Schedule for Representative Station Design/Construction Project

Stage	Time frame
Appointment of Building Committee	1 month
Selection of Architect	4 months
Preliminary Design	3 months
Commitment of Funding	3 months
Construction Documents	6 months
Local Regulatory Review	3 months
Bidding	1 month
Award of Contracts	1 month
Contractor Mobilization	1 month
Construction	12 months
Move-in and Occupancy	1 month
Contract Close-out	3 months

The actual design of a station or facility takes place once the need for a station has been identified and established by the department's hierarchy. The first step in this process is to appoint a building committee. The building committee is generally appointed to determine many of the details required in developing a facility design and to oversee the design/construction process. The actual design process is carried out by the design team which begins once an architect is chosen and consists of the architect and other experts in building design.

In isolating the design part of the construction process, it is important to understand how the department, the building committee, and the design team interact. These interrelationships, illustrated in Figure 2, show how each group works together for designing the station:

1. The ***department administration*** or governing authority decides on the need for a station based on department operations and community expectations. A department official, such as the department chief or a member of his or her staff, appoints a building committee.

2. The ***building committee*** may conduct all or part of the needs assessment to determine requirements for the station design. The building committee's primary responsibility is to oversee both the design and construction process by interacting with the design team and contractor.

3. The ***design team*** prepares the detailed station design based on the requirements of the department and through interaction with the department administration and/or building committee. The design team translates the department needs and requirements into a specification from which the contractor can build the station.

While the above descriptions represent one possible relationship between the groups involved in designing the station, there are several variations of this process. For example, the degree for which the department defines its needs can be split between the department administration or building committee. In some cases, departments may depend on outside specialists or consultants. Larger departments or departments which are part of local government may have groups (i.e., standing building committees) already established with these responsibilities.

The specific process for how fire/EMS departments, local governments (or other department authorities), architects, and construction engineering firms interact to develop a specific station design varies, especially in the consideration of safety and health regulations, building codes, and other specifications. An example of this process is embodied by the following steps:

1. Determination of fire/EMS department and city needs;

2. Determination of project constraints (e.g., location and safety/health restrictions);

3. Development of the preliminary station layout;

4. Determination of the type of team needed to carry out the project (the types of groups to be involved);

5. Determining financing constraints; and

6. Implementation of the development plan.

Of course, depending on the groups involved and the formality of each group's involvement, this process can vary tremendously, Nevertheless, the following guidelines apply:

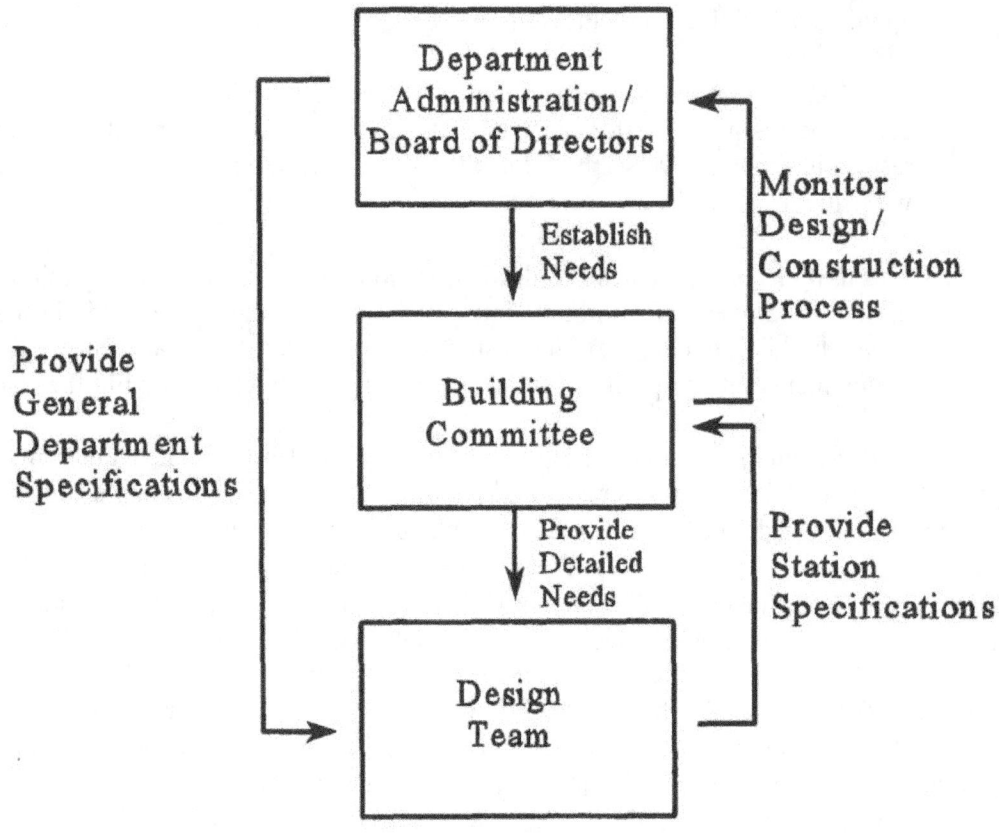

Figure 2. Interaction of Department, Building Committee, and Design Team in Station Design Process

- In all cases, the design process should start with a **needs assessment** of the community, the organization, and their expectations.

- Fire or EMS department administrators should endeavor to provide its citizens the best facility for their available budget in terms of meeting the priorities established in the needs assessment, particularly those related to safety and health.

- The location of the structure in consideration to the community's growth and response time plus the enhancement of total coverage with existing or neighboring stations is a paramount feature in new station's design.

- Initial layouts should be proposed based on previous successful designs or as affected by factors for the facility's location and function.

- Details should be handled by a team of individuals representing various areas or expertise as related to the design project, The team should include at least one individual who is knowledgeable in safety and health regulations or who at least has access to these documents.

- Financial constraints should be determined and compared with preliminary estimates of total development costs.

- The actual developmental plan should encompass all of these elements with the objective of completing a design which meets all identified needs and constraints.

The planning phase of fire stations is a complicated task and requires close cooperation between all parties involved, particularly in the early stages where department station needs are defined.

The Needs Assessment

The first consideration of station planning should be a **needs assessment** of the community and department for listing specific requirements for the station in terms of:

- station capabilities,
- equipment (i.e., apparatus) accommodation, and
- personnel safety and health.

The station needs assessment should begin with community expectations and operational needs which can be broken down into more specific requirements. For example, a general need might be to accommodate an engine company and EMS company at the same station. A specific need within this general need is to provide bay capable of housing two apparatus and EMS vehicle. As will be explained in later sections, there are several factors to be considered. The building committee or other responsible group must clearly define these needs and where possible quantify their requirements so that proposed solutions allow measurement.

There are several valuable sources of information which should be consulted during the process of defining specific needs for the design and construction of a station, including:

- the department's own history with existing stations (a review of existing stations' capabilities, equipment accommodation, and personnel safety and health can point to aspects of the design which work and those which do not);

- seeking advice from neighboring departments;

- examining handbooks, articles, and guides (in addition to this manual) for ideas (one of the best references that provides some depth on this subject is the Fire Chief's Handbook[14]); and

- hiring a consultant who specialized in fire/EMS facility planning and design,

The needs assessment should include **assessment of risk** which identify safety and health requirements. The appendix of NFPA 1500, *Standard on Fire Department Occupational Safety and Health Program,* provides a series of recommendations for preparing a risk management plan. The risk management plan includes:

- the identification of risks,
- evaluation of those risks,
- methods to control risks, and
- monitoring risks through follow-up actions.

Table 3 provides a summary of recommendations contained within NFPA 1500. These recommendations should be followed by the building committee in their process of deciding how to address safety and health requirements.

The specific identification of risks and the evaluation of their severity may be best examined using a visual model. This type of model examines inputs to the station and outputs from the station which may constitute some risk. Figure 3 provides an example of a simple graphic analysis of factors which impact the safety and health of personnel at the station. While it may be used as a general model, individual departments or building committees are encouraged to carefully investigate the particular concerns of their local area and department.

It is important that the safety features be a principle part of the design process. Lack of these features will cause problems later on and that old saying of *"pay me now or pay me later"* will be a living testimonial. Examples of important safety features include:

- drive-through features (to prevent vehicle accidents),
- guarded rails and single floor layouts (to mitigate falls),
- large apparatus bays (to provide adequate personnel access),
- control of diesel emissions (to reduce chemical exposure), and
- compliance to safety codes (to mitigate other hazards).

[14]Paul de Silva, "Fire Station and Facility Design," Chapter 14 in *The Fire Chief's Handbook,* 5th Ed., J. R. Brachtler and T. F. Brennan, Eds., Fire Engineering, New York, 1993, pp. 477-516.

**Table 3. Steps in the Development of a Risk Management Plan
(Recommended by NFPA 1500 in Paragraph A-2-2.3)**

1. *Risk identification.* For every aspect of the operation of the fire department at the station, list potential problems. The following are examples of sources of information that may be useful in the process:

   a. A list of the risks to which members are or may be exposed.

   b. Records of previous accidents, illnesses, and injuries (both locally and nationally).

   c. Facility and apparatus survey/inspection results.

2. *Risk Evaluation.* Evaluate each item listed in the risk identification process using the following two questions:

   a. What is the potential frequency of occurrence?

   b. What are the potential severity and expense of its occurrence?

   Use this information to set priorities in the control plan (needs assessment). Some sources of information include:

   - Safety audits and inspection reports.
   - Prior accident, illness, and injury statistics.
   - Application of national data to local circumstances.
   - Professional judgement in evaluating risks unique to the jurisdiction.

3. *Risk Control.* Once the risks are identified and evaluated, determine which control should be implemented and documented. The two primary methods of controlling risk, in order of preference, are:

   a. Wherever possible, totally eliminate/avoid the risk or the activity that presents the risk. For example, if the risk is falling on ice, then do not allow member to go outside when icy conditions are present.

   b. Where it is not possible or practical to avoid or eliminate the risk, take steps to control it. In the example above, methods of control would be applying sand/salt or wearing of proper footwear.

   Also consider the specific development of safety programs, standard operating procedures, training and inspections as control methods.

4. *Risk Management Monitoring and Follow-Up.* Periodically evaluate the selected controls to determine if they are working satisfactorily. If not, identify and implement new control measures.

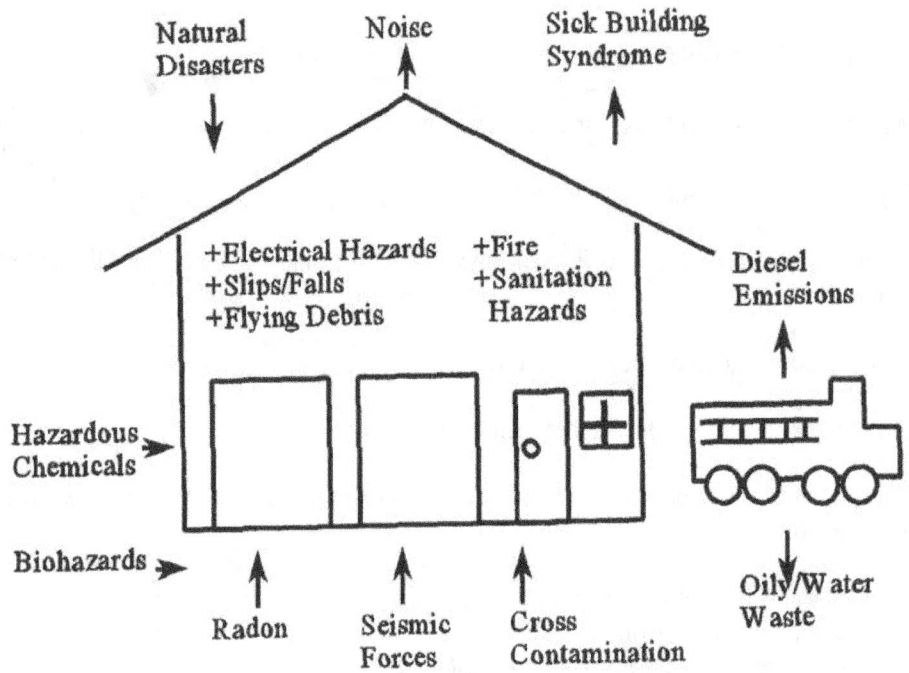

Figure 3. Graphical Representation for Identifying Station Hazards

Once department needs have been projected and regulatory needs are identified, the assistance of an architectural firm (or design team) can be obtained. The design team will provide the department with preliminary drawings for reviewing department concepts and compliance with regulatory agencies. In many cases, the building committee serves as an interface between department hierarchy and the design team.

The Building Committee

As the interface between the department and the design team, the building committee serves an important function in ensuring that department needs are met. The degree of autonomy provided to the building committee will depend on the individual department. In addition to supervising the design/construction process, building committee responsibilities can include:

- completing the needs assessment,
- choosing the architect,
- monitoring the project's schedule and budget, and
- making recommendations or decisions for approving changes.

The design or redesign of existing facilities can have several different groups involved in the building committee providing representation for various groups, including:

- department administration or staff,
- line personnel (emergency response personnel),
- labor (if department personnel are represented by a union),
- budget or other city/county/area officials controlling financing[15],
- public works department representatives,
- safety or code and regulation personnel,
- citizen groups, and
- special consultants.

Each of these groups has its own perspective on the design of a station and will want its own specific needs addressed. The constraint faced by each group is the limitation of funds available for building the station, Therefore, the attainment of each group's objectives is often a function of how these funds are allocated for station's construction. Table 3 describes the typical perspectives for each of these groups.

Ideally the department should choose qualified individuals to serve on the building committee. While some representatives are likely to be selected owing to their current positions with the department or other part of the community government (for non-volunteer departments), departments should assign individuals who have some expertise in construction principles, designs, techniques, and budgets, since changes identified early in the process are much easier to correct for.

---

[15]Many volunteer fire departments use a governing body such as a Board of Trustees or appointed/elected groups (e.g., commissions) for governing major purchases by the fire department.

Table 4. Roles and Objectives of Groups Represented on the Building Committee

The *department administration and staff* should be able to provide a historical overview of what has worked and what hasn't worked for the department. They will also be keenly aware of how the station must function to meet overall department operational needs.

Line personnel who must live and work in these facilities should have an active voice in the design process and will provide the most beneficial information. These individuals know what it takes to make the structure work both functionally and comfortably. Studies have proven that firefighter comfort greatly influences their productivity. These personnel are often the individuals who are quick to point out deficiencies.

Labor representatives will also speak for the line personnel and try to obtain as many features as possible for increasing their safety and comfort.

Budget personnel clearly want the most economical structure for meeting their budget. They may not have projected funds to cover features based on requirements from standards that are voluntary. Arguments should be made to justify these additional requirements through showing improved safety and decreased department costs (in a cost-benefit analysis). Failure to comply with existing federal, state, or local requirements during the initial planning process will usually add greater costs later when modifications are required. Therefore, part of the cost-benefit analysis should show how higher payouts in future years will occur if certain requirements are overlooked in the initial design process.

Public Works representatives (for local government stations) will examine station design in terms of longevity and maintenance concerns.

Safety Office Representatives should be involved during a fire/EMS station design process. The local authority usually asks that the structure meet all applicable codes and ordinances, typically those found in the uniform building and fire codes. They may not consider all federal, state, or consensus standards that are outside the scope of building or fire codes. As most consensus standards like NFPA are voluntary, there has to be a clear need demonstrated by the fire service to meet these requirements (by performing cost-benefit analyses or by showing increased department liability).

Citizens groups will want to ensure that the station does not affect the local community in detrimental ways such as providing pedestrian safety hazards, noise, or an appearance inconsistent with the aesthetics of the neighborhood. Their participation in the building committee should be considered.

Consultants may be hired for any number of reasons to assist during the design process with some special need or provide a general understanding to help departments who are not familiar with particulars of the station design process.

The Design Team

The design team will consist of the architect and other experts who translate specific emergency response department station needs and requirements into a set of plans from which a contractor can build the station. In most cases, the design team will at least include architect and other design specialists from outside the department, but it is possible that some larger organizations will have an in-house facilities design staff with their own architects and other experts.

Depending on the size of the architectural firm and the resources available to the department, several specialists may be required representing different areas of expertise. In addition to the architect, station design can conceivably require a design team where the specialties are listed in Table 5 are represented.

Table 5. Design Team Specialty Areas

● civil engineering	● audio-visual
● mechanical engineering	● hardware
● structural engineering	● space planning
● plumbing engineering	● interior design
● electrical engineering	● industrial hygiene/safety
● acoustical engineering	● infection control
● lighting	● community relations

Larger architectural firms will already have specialists in many of these areas. Other architects may be required to hire advisors representing these areas. In large organizations, the department may choose to provide design expertise in some of these areas. Whichever means is used to assemble a design team, it is important that all areas of expertise be represented.

The choice of an architect is of critical importance. If the department has previously hired a designer which has worked well in the past, then their services should be considered. Likewise, architects with no experience in fire station design should be avoided. Good architects are trained to reconcile a department's wish list with the realities of the construction budget and translate the result into a set of plans and specifications that the contractor can easily use. Architects are charged with preparing designs which meet the organizations' specifications while using their skills for making the building blend in with the community. In some cases, the architect may add characteristics to the structure which can be appealing to other potential clients but might not be needed by the department and this additions can become the source of future problems. Overall, the department must feel comfortable in working with the selected architect and have confidence through some demonstrated experience. Table 6 provides guidance for choosing an architect or architectural design firm.

```
┌─────────────────────────────────────────────────────────────────────────┐
│                                                                           │
│              Table 6. Guidelines for Choosing An Architect                │
│                                                                           │
│                                                                           │
│   The selected architect should have:                                     │
│                                                                           │
│   1.    Experience in the design of public or industrial facilities,      │
│         specifically fire or EMS department stations;                     │
│                                                                           │
│   2.    Demonstrated successful performance in the design of fire or      │
│         EMS department stations;                                          │
│                                                                           │
│   3.    Familiarity with government contracts (for local government-      │
│         based departments);                                               │
│                                                                           │
│   4.    American Institute of Architects (AIA) certification and          │
│         state/local licenses for working in the area of architecture;     │
│                                                                           │
│   5.    Design team members who are equally qualified within their own    │
│         respective fields of expertise;                                   │
│                                                                           │
│   6.    Recommendations from other departments or organizations with      │
│         similar operational needs.                                        │
│                                                                           │
└─────────────────────────────────────────────────────────────────────────┘
```

Even though the architect will carry the primary responsibility for the station design, there are several other entities, not previously mentioned, that need involvement. These typically include other parts of the city or county organization such as the legal department, accounting office, equal employment office, health department, and the office of construction and land use. In the case of volunteer fire departments, this may include the board of trustees or community representatives. When each of these groups can cohesively work together, the acceptance of the project is enhanced, the project remains within budget, and the community and department are able to meet their expectations.

SECTION 3 - OVERVIEW OF DESIGN CONSIDERATIONS

The design and construction of a fire or EMS station is a long and complicated process. A number of different decisions are required and different persons may be involved in those decisions. The factors governing station or facility design vary from department to department and even within the organization itself. For some departments, the construction of a station may be an on-going process as the community or area which it serves is continually growing. These departments may have in-place building committees and well-defined procedures and specifications. In other organizations, the building of a station may be an infrequent event, requiring new research each time due to changing topography, demographics, and industrial developments. In all cases, facilities should be designed which account for all relevant regulations and promote an appropriate level of safety and health for the occupants.

This section provides an overview of the various considerations which go into developing the appropriate design. While many design choices are based on organization's particular needs for the new or remodeled facility, there are a number of design options which are regulated and impact personnel safety and health. The following sections provide insight for considering safety and health issues when making design choices.

Design Philosophy for Fire and EMS Stations

The old National Board of Fire Underwriters followed by the American Insurance Association originally set the tone for fire station design. Their concerns covered such factors as quick, efficient, and safe response of apparatus to alarms. EMS responses and exposures were not found until departments accepted these responsibilities. With time, fire departments began to become more involved with prevention activities rather than exclusively suppression functions. The fire station began to take on the duties of a business. Stations included administrative offices for the various functions of the organization with fire prevention and education was conducted at these facilities. Firefighters and EMS personnel no longer went to the fire house to be trained by their peers but instead needed specialized training and training facilities. Changing functions for the fire station meant drafting pits, hose towers, classrooms with audio-visual support, security, and varying community expectations.

As the fire and EMS departments' attention turns to the flexibility of the station, several questions are raised:

- Can the station be used for other municipal functions?

- Can department needs be met with shared facilities?

- How long should these facilities last?

- What are the most cost effective way to build these facilities?

- What should be the size and spacing of stations?

19

- What are the standards and regulations that must be considered in the design and construction process?

- How can the department maintain its present level of service and be ready to grow in the future?

Design Factors

Fire and emergency medical services station designs have changed over the past decades. They are being recognized more and more as specialized facilities with their own specific design approaches. Design and facility features differ among fire stations, EMS stations, and other types of facilities based on a variety of factors, including:

- the role of the station,
- the type of department and expected response level,
- specific functions of the facility,
- integration with joint or shared facilities,
- community restrictions,
- future requirements, and
- available resources.

These same factors also can impact the level of safety and health for personnel working and living at these stations.

Role. NFPA 1201, *Standard on Development of Fire Protection Services for the Public,* establishes the primary functions of the fire station as:

- fire prevention and risk reduction,
- fire suppression,
- rescue and emergency medical services,
- hazardous materials response, and
- disaster planning.

Differences in station design can occur as a result of slightly different roles of the specific department. Departments with integrated emergency medical responsibilities have additional station design requirements for accommodating EMS needs such as shown in Table 7.

While these functions represent the more traditional purposes of fire and EMS station, other functions of a fire or EMS stations could provide for:

- *Civic activities* - Fire stations are community assets and often part of the local government. A common civic function for the fire station is to serve as a polling place.

- *Social activities* - Many fire stations also provide meeting rooms and other facilities for community-based groups. In some communities, the fire station may serve volunteer social functions. Stations also have frequent tours.

Table 7. Impact of EMS on Station Design

- special storage requirements (e.g., secure spaces for drugs, refrigerated medical supplies);

- routine handling of medical gases (oxygen);

- segregation of station alarms (fire versus EMS);

- differences in shift length/separate quarters; and

- differences in vehicle requirements.

Key safety and health concern for these functions are often related to the number of people that the facility must accommodate. Staffing levels and number of personnel on duty are important. The total number of people that can be in the facility at any one time, due to training, a disaster, or community event, should be considered.

Type of Department and Expected Response Level. There are differences in station design whether the facility is in a metropolitan area or a rural area, whether fully staffed by paid personnel or partially staffed by volunteers, and whether headquarters or substation or having specialized functions. Some of aspects of station design are also affected by the overall response level as some stations are required to do multiple runs in the immediate local area of the community, while other stations may have infrequent demand but cover relatively larger areas, Examples for the impact of department type and response level on station design include:

- Paid departments must be based on the assumption that firefighters or EMS personnel will be responding from the day room, offices, or sleeping quarters to the apparatus. This consideration dictates the layout of the facility in terms of apparatus bay accessibility to these areas. However, at the same time, some separation is required to prevent fumes and noise from reaching sleeping areas and other interior living spaces.

- Volunteer departments are more apt to be less formal and more group oriented. Sleeping quarters may not be needed but these stations often require ample parking for effective response and building security measures.

- Combination departments require flexible facility designs to accommodate the needs for both volunteer and career firefighters and EMS personnel).

Questions addressing these issues include:

- Is it a volunteer department station that is to be unmanned for long periods of time, a volunteer department station with individuals in and out of it daily, a part-time paid department, or a paid department?

- Will the department be located in a rural area, industrial area, commercial district or residential community? Each of these sites and applications require different design approaches. Furthermore, each of these types of departments may require a different source of funding to complete the task.

Specific Functions. Once the type of department is known and the community it serves is defined, the specific functions for the facility must be determined. A rural department with residential and small commercial occupancies would not consider aerial ladder apparatus or similar vehicles in its station design and planning. If EMS service is to be provided, space must be allotted for decontamination facilities for the equipment and personnel. NFPA has and local health departments may have requirements for features within these type of structures, It may also be important to consult with the local area hospital for compatibility between department and hospitals to enhance the responsiveness of both facilities. If the station is permanent staffed, then station occupants comforts should be addressed. If the station will be used as a community gathering place, consideration should be given to provide a separation between the functional part of the station and the civic side of the facility. The department administration should decide how long the station is to last, what its growth will be, how to use the facility to meet the communities needs, and how to look after its occupants.

Other functional assignments of the stations will affect facility design. Headquarters units may require central offices, dispatch facilities, and training areas. Some facilities may also be a site for apparatus maintenance, marine units/fire boat, or other special functions. There are often different safety and health concerns for these special functions.

Local Geography. Geography and climate affect several aspects of station design in terms of safety and health. Examples include:

- roof design for local weather conditions (wind/snow loads),
- choice of substructure and structural elements to withstand seismic activity,
- terrain for actual facility layout, particularly driveways and aprons,
- exterior/interior drainage which prevents ice buildup during winter months,
- adjacent roadway visibility for apparatus access, and
- overall space for accommodating the entire station layout (some locations accommodate drive-through apparatus bays, others do not).

Most of these aspects will be specific to the area and usually within the common knowledge of the architect, but only if the architect is made aware of the special needs that fire and emergency medical services have compared to regular industrial facilities.

Community Restrictions. Community restrictions can take two forms:

1. The community may have its own building codes and regulations to which the station must adhere; and

2. The community may have certain expectations that the facility fits with the rest of the architecture in the area or at least does not interfere negatively with the adjoining community.

Many communities have specific laws governing use of utilities, construction materials, and right-of-way. It is impossible to list the various restrictions which can be imposed. Nevertheless, this aspect of station design is usually handled in the review of building permits where proposed designs and layouts are reviewed against existing local codes, zoning, and ordinances. Some communities require that the fire station appear to blend with other structures in the community. In these cases, the appearance of the station may be critical from an aethestic point of view (Figure 3 shows photographs of two fire stations, one based on a traditional appearance in an urban setting, the other of modern, rural design).

Fire department officials should meet with community leaders to discuss the station's design and solicit their input. The neighborhood should be sought as an ally to work with to avoid future problems. While this practice may only affect certain elements of the station's exterior design, some requirements can affect station layout and overall safety. Commonly, facilities are required to shield neighboring structures from noise by use of shrubbery and response functions should be tied into local traffic lights.

Flexibility for Future Requirements. Facility requirements change with time. This can occur as response needs change or as department functions are added or deleted. What is often built to meet the department's current needs, requires periodic reevaluation and possible upgrading to meet the demands of the future. As the fire and emergency medical services grow and expand, new facilities are needed for specialized training or testing of equipment. Flexibility in the station's design may provide for:

*   special training features (such as training towers),
*   hazardous material decontamination sites for gross decontamination,
*   medical disinfection sites,
*   commissaries (for distribution of uniforms and personnel protective equipment),
*   accommodation of female emergency responders,
*   addition of more personnel and apparatus, or
*   consolidation with EMS and law enforcement agencies.

In order to meet these challenges, some structures must be redesigned accordingly.

Separate facilities might be designated to have capabilities for specialized training and activities including:

*   confined space,
*   high angle rescue,
*   auto extrication,
*   pumper testing,

Figure 3a. Fire Station with Traditional Appearance in Urban Setting

Figure 3b. Modern Station Design for Rural Location

- ladder testing, or
- self contained breathing apparatus maintenance.

This practice enables departments to evenly spread resources while having qualified individuals in all areas.

In addition, departments must consider how the fire service will change in the long term future. Just as the incorporation of EMS represented a major change in fire department responsibilities and impacts on station requirements, so should other areas which some fire departments are currently considering (see Table 8).

Table 8. Possible Future Roles of the Fire Service and Their Consequences

Possible Future Role	Consequences for Station Design
Future diversity of services	Increased specialization and equipment needs (requiring additional space storage), and utility needs) communications;
Public health prevention/immunization center	Increased public access, staffing, drug storage and handling
Place of shelter and safe house	Increased public access, staffing, security
Hazardous material control	Special chemical storage spaces

Resources. Every organization would like to build a station which possesses the state-of-the-art equipment and uses the best materials. Unfortunately, most organizations are constrained by limited resources and must often make tradeoffs between desired feature and practical design needs. Nevertheless, these choices should not compromise personnel safety and must obviously be in full compliance with all applicable codes and regulations. Failure to meet these requirements often results in expensive redesigns and remodeling. Even though it is difficult to justify, it is also important for design of stations and other facilities to anticipate future requirements. These requirements are often in the form of increased functionality or accommodation of new equipment. However, station designs should also forecast new regulations as new safety concerns are identified. For example, the American Disabilities Act, created the need for stations across the U.S. to be modified.

Siting and Space Planning

Site Selection. The former American Insurance Association Special Interest Bulletin (which is no longer printed) covered some of the factors which should be considered when constructing new stations, including:

- response time,
- area of response coverage, and
- ability to concentrate department resources.

In addition, site selection should accommodate:

- station areas for future growth or storage,
- locating fuel sites in the event of disasters,
- emergency power needs in the event of prolonged power outages,
- apparatus response in more than two directions,
- ample parking for personnel working at the facility,
- training needs,
- communications considerations (including line-of-sight), and
- station distance from the curb to avoid potential pedestrian or vehicular accidents.

All site selections should include a Phase I Environmental Site Assessment which encompasses a thorough review of the sites prior history and potential contamination.

Layout. The basic layout of the station has several different possibilities. In terms of safety and health, principal concerns which arise in selecting the layout are:

- apparatus driveways,
- internal station traffic, and
- proximity of disinfection/washing areas.

Stations designed with drive-through features are usually less likely to have vehicular problems (see Figure 4). There should be several means of exiting from the business side of the structure to the apparatus bay in order to respond quickly. Where interior doors are provided they should have a safety glass panel to enable individuals to see what's on the other side to avoid potential collisions. If sliding poles are used, they should have cushioned mats at the bottom. The entry into the pole area should also be protected against falls through the opening and objects that could roll along the floor and fall through the opening.

Utilization of Space. The amount of space in a station should always account for some growth and if possible allow for more apparatus than initially intended. Some space should be allotted for addressing specific health and safety requirements. Construction costs go up each year and it is far easier and cheaper to over-build than to go back and get the financing for remodeling. For example:

- Gender-specific facilities should be considered even if it is not a current concern at the facility.

- Separate, controlled spaces should be provided for storage of batteries, hazardous chemical storage, and personnel protective equipment (storage space is nearly always underestimated).

- Space should be allowed for expansion or changing department needs.

Space requirements should take into consideration the size of the lot, the anticipated use of the structure for other municipal functions, and other possible use for best utilization of the structure.

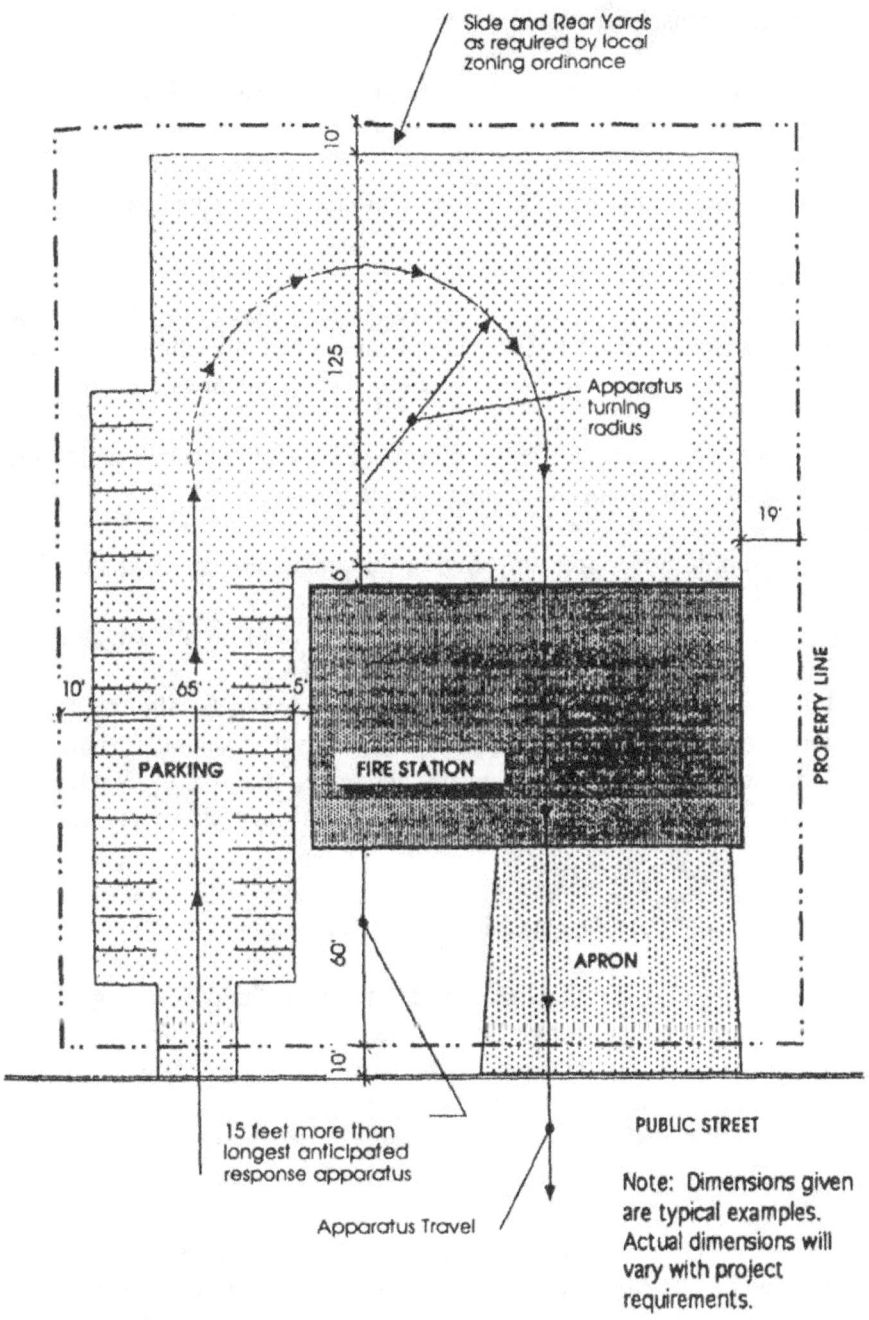

Figure 4. Example of Station Design with Apparatus Drive Through Layout

27

Exterior Design Considerations

Foundations. The soils are the starting point for any new design and construction process. Pertinent considerations include:

- Evaluation of soil conditions for stability and seismic elements factored into the footings;

- Foundations to take the heavy load of today's heavier apparatus; and

- Ensuring adequate safety factors for the station in either good or bad soil conditions. With many new structures, casing holes are drilled to solid ground and then filled with concrete. These casing holes are often reinforced with rebar for strength. Similar methods are used for remodeling.

Floor Slabs. Stations floors are required to meet the local building codes and should be substantially reinforced in apparatus areas. In addition:

- Floors should be sloped and provide for drainage to prevent standing water.

- The slope of the slab should not allow outside water to run back into the structure.

- Waste oil products from these apparatus should not be allowed to flow into city storm sewers. Some structures use oil and water separators built into the floor to deal with waste residue from the apparatus, others use a simple sump pit and have it pumped out when full. Concrete floors also need to be sealed for porosity.

- Apparatus floor slabs should be finished to become slip-resistant.

- Safety stripes should be painted on the floor for the positioning of the apparatus. For further safety in stations housing aerial apparatus, a marker should be painted on the apron of the station or in the street denoting when these apparatus may start a turn when leaving quarters.

- In stations using hose dryers, additional load factors should be considered.

Structural Systems. The structural system will usually be either a frame system where a steel, concrete, or wood skeleton supports the building, or a load bearing wall system where the walls of the building, usually masonry, support the roof and second floor framing. Often a combination of the two systems is used. The choice is usually based largely on economics and then the needs of the building. The larger the building, the more likely it is that a frame system should be used. Steel, poured concrete, precast concrete, or heavy wood timbers can be used for frame. Support floors and roofs can be precast concrete plank, poured concrete slabs on steel joist, metal decks, or plywood on wood joists. The selection of wall, floor, and roof structures are based on local practices, the availability of materials, and particular structural concerns (like seismic activity). From a safety and health perspective:

- The frame should be designed to eliminate columns from apparatus bays since columns are both hazards to the apparatus and to personnel responding to the apparatus (columns can be eliminated by increasing the size of the framing members; this increases the cost of the building but it is a worthwhile investment in terms of safety and future adaptability).

- Consideration should be given to reinforced flooring and roof structures. Reinforced which are more likely to resist collapse during adverse weather conditions and earthquakes.

Walls. Walls can be either load-bearing or non load-bearing. Load-bearing walls usually are limited to masonry construction, brick, concrete block, or both. Non-bearing walls can be masonry, or precast concrete, metal panels, or any number of proprietary system. The side and upper surfaces of the parapet wall (roof level) must be protected in areas where ladder drills will be carried out-front, top, and rear. Different types of wall construction methods afford different advantages and disadvantages:

- Generally masonry or precast concrete represent the most durable cost effective, low maintenance wall system, but they must be designed and constructed carefully for proper performance. Both brick and block are available in a large variety of colors and textures. Precast concrete can be fabricated to almost any shape, texture, and color. Its cost effectiveness varies from region to region depending on transportation cost. Natural stone veneers are durable but expensive.

- Metal panels can be inexpensive or very expensive depending on the design and material selected. The joints have a tendency to leak, however, and less expensive systems deteriorate rapidly.

- Exterior Finish and Insulation Systems (EFIS) represent one of the fastest growing new technologies in construction. They consist of a thin coating of synthetic acrylic or epoxy modified stucco applied to insulation bound over wood/metal studs or a masonry back up. EFIS offer ease of construction combined with good insulation and highly decorative surfaces at very little cost. However, they may not be durable and may be easily damaged, and are combustible. They should be avoided in areas subject to abuse or vandalism.

Roofs. Two types of roofs can be used-flat or pitched, representing differences in economy and performance, weather, and seasonal conditions.

- Flat roofs are pitched for drainage at about 1/8 to 2 inches per foot. Flat roofs are considered more economical because less materials are required in their construction. However, this construction increases the possibility that water will remain on the roof and find a hole or imperfection to penetrate. If appropriate materials are used and if properly constructed, a flat roof will perform as well as a pitched roof. In geographical areas of the country where rain is prevalent, flat roofs generally do not provide adequate runoff. Flat roofs should be avoided in areas of the country where snow is a major factor because of the weight buildup.

- Pitched roofs have slopes with 2 inches of rise for each foot of run or are steeper. These roofs can be covered with asphalt shingles, slate, tile, or metal systems, Asphalt shingles are the least expensive and are available in multiple textures and colors which can be expected to last 25 years or more. Slate and tile are more durable but more expensive and harder to maintain. Metal systems require careful design and construction; in fact, the less expensive types should be avoided.

The Fire Chief's Handbook recommends that the selected roofing system have a UL class A fire rating and a UL class 90 wind uplift rating where high winds are a concern.

Windows. Window construction materials include wood, steel, aluminum or vinyl in custom or standard sizes. Construction cost and insulation performance of these window types vary.

- Wood frame windows are considered strong, light weight, and energy efficient, When the wood is protected by some other material (such as metal or a heavy coating), durability is improved.

- Aluminum and steel windows in the heavy commercial grades are sturdy products intended for hard usage. However, cheaper grades metal windows offer poor performance. These windows require special thermal break construction to obtain the same energy efficiency inherent in wood.

- Vinyl windows are the least expensive. Since this type of window material is relatively new, their long term performance is unknown.

The selected glass work or glazing should be high performance, low-E insulated units with an R factor of 2.5 or better in all but the mildest climates. The slight increase in cost is easily justified in energy savings and personnel comfort. In high vandalism areas, polycarbonate, an impact-resistant plastic, should be considered. Alternatively, window screens or protective shutters can be used. The likelihood of glass breakage should also be factored into the selection of windows. Where fire rated glazing is required, wire glass is most commonly used. New technology fire-rated glass, gel, and ceramic composites which eliminate the visual distraction of the wire and have recently become available but they are much more expensive. Window designs which permit easier cleaning such as double hung vertical windows should be considered to lessen to possibility of falls from cleaning.

Life Safety Code requires that sleeping areas have a minimum size window, usually 90 inches, as measured by (1) vertical length, and (2) horizontal length, to accommodate emergency rescue.

Doors. Two types of doors warrant consideration from a safety and health perspective-exterior doors and apparatus doors.

Exterior doors are usually steel "hollow metal" or aluminum and glass. Wood doors can be used but are more susceptible to weathering and break-in. Proper

30

weather stripping is important in cold climates. Hardware should be rust resistant, accessible, and suitable for heavy usage. Hollow metal doors should be galvanized. All doors leading onto the apparatus floor should swing toward the apparatus. Furthermore, overhead doors must be controlled from both the watch office and at conveniently located floor stations preferably right next to the primary path into the bay. Any doors with glass should have safety glass or wire-reinforced glass, particularly for those doors which enter into the apparatus bay.

● Apparatus bay doors can be coiling or sectional types. Coiling doors are more expensive and generally slower. Sectional doors are most common. Care should be taken to avoid large doors that accommodate two pieces of equipment responding out of the bay. These door are heavy and subject to more failure. Doors should be large enough to accommodate the largest apparatus in the department's fleet. Wood doors are acceptable in milder climates and are easier and less expensive to repair when damaged. In colder climates, doors must be insulated but insulated wooden doors often warp and delaminate. Insulated and weather stripped steel or aluminum doors are best for these situations. Electric door controls should also be used.

For durability and safety, the Fire Chief's Handbook recommends that apparatus doors include:

● use of heavy duty geared motor operators,
● high-cycle six inch torsion springs,
● properly anchored and supported tracks,
● safety sensor strips on the bottom edge of the door that automatically stops or reverses door movement when the door comes in contact with personnel or the apparatus, and
● local and remote controls, including a radio control in the apparatus.

Some station designs use magnetometers in conjunction with automated dispatch systems. These systems are designed to provide an indication when an apparatus leaves the station and may be used as automatic door closing devices for stations that have swinging doors rather than overhead doors. There have been a number of reports that these systems have closed the doors on the apparatus as it was leaving the station. Furthermore, expensive maintenance costs appear to be associated with this type of system. One source suggest placing a pressure-activated strip across the door floor opening to provide the same capability at a more reasonable expense. Use of special lights indicating the position of the door has also been reported as being useful.

The American Disabilities Act (ADA) requires that some entry/exit doors be power-assisted. In addition, entry/exit doors for the station must be fire-rated, usually for two hours (check with local authority for specific required fire ratings).

Station Parking, Ramps, and Exterior Driveways. The primary areas directly outside the station must be properly design to reduce likelihood or accidents and provide for efficient operations, Considerations include:

- Providing sufficient, secure parking space which can accommodate shift changes, does not interfere with the drill court capabilities of the station. The station should be designed to accommodate the parking of reserve apparatus inside. It is a good idea to install "NO PARKING" signs on both sides of adjacent streets, to permit unobstructed response, in any direction. Curb areas which are hazardous should be marked.

- Ensuring adequate lighting in parking areas and walkways (street and exterior structure lights) at the station for improving security and providing for pedestrian safety at night.

- Constructing station ramps and drill courts with concrete. Asphalt will not withstand the heavy weight of fire apparatus or the sharp spurs of heavy ground ladders. Furthermore, asphalt softens during the hot weather, compounding these deficiencies and patching asphalt creates uneven surfaces.

- Designing station courts and ramps, as required by some local codes, so that water that accumulates during drills does not run off into adjacent properties. In a similar manner, station entrances should be designed to prevent water seepage under the large apparatus doors.

- Complying with the American Disabilities Act (ADA) as fire stations are public buildings and often times used as polling places. This may require the use of ramps or lifts depending on the configuration of the facility.

Other Exterior Considerations. Several other considerations are important in the design for establishing appropriate levels of safety and health. Among these are:

- Planning the side yards of stations to be of sufficient width to accommodate long hose lays during drills and hose testing which will not encumber adjacent sidewalks or create hazards for passing pedestrians.

- Connecting the alerting system for the station to alarm or speakers at both the front and rear of the station to alert personnel working in these areas. These speakers should be silenced during the evening hours which can be achieved by an on/off switch located on the watch office console (NOTE: The alerting system should not be confused with the station fire/evacuation alarm which must remain active at all times).

- Complying with fire and building codes which require fueling pumps to be located outside station facilities.

- Selecting frost-proof gate valves (hose bibs) on the exterior wall of the drill court and at the front and side of the structure which are sized for 1½ inches. If the station does not have a hose tower, a dummy standpipe Siamese should be installed for drill purposes. Additionally, a hydrant needs to be installed on the drill court to facilitate realistic training.

- Electrical overhead connectsions within proximity of water hoses need to be rated for damp/wet locations.

- Providing for filling booster tanks; An economical way is from a water source on the station exterior or by stretching a hose line across the apparatus floor. The Fire Chief's Handbook recommends overhead lines in the apparatus bays. They can be simple hose drops at each apparatus position or they can be mounted on centralized electrically powered reels. One and a half-inch hose with a trigger nozzle similar to a gasoline pump works well. If reels are used they must be specially equipped with hose guides and clutches to feed from the vertical position. A 40 foot hose mounted on a reel between two double bays can serve four apparatus positions.

- Furnishing facilities where drafting pits are used with the necessary guards and markings that indicate a DANGER area.

- Equipping the stations with a covered public telephone or intercom near the entrance that goes to the alarm office in the event the fire station is unoccupied.

- Accessibility of garbage cans or dumpster to the garbage collection trucks which are located off the drill court.

Interior Design Considerations

General. The American Disabilities Act (ADA) requires minimum hallway widths, hallway design free of protruding objects, specific water fountain heights, and special fixtures and dimensions for bathroom facilities. All facilities should provide for adequate evacuation and emergency egress as defined in NFPA 101, *Life Safety Code*.

Walls. The principal concern for interior walls are their fire resistances and durability, Masonry-based construction should be used in areas subject to rough use such as the apparatus bays. Gypsum board on metal studs is common for other spaces. Finishes need to be appropriate for the usage of the particular space:

- areas of rough usage should include a spray glaze or high-gloss paint,
- in washroom and laundry areas, ceramic tile should be used, and
- semigloss paint or vinyl wall covering can be used in other areas.

Flat paint is usually not appropriate, except possibly in office areas. Wood paneling should only be used if it has the proper flame spread rating.

Ceilings. Acoustical types of ceiling enhance noise reduction. Acoustic ceilings come in a wide variety of styles and sizes (e.g., 2x2 and 2x4 panels). Concealed spline ceilings and gypsum board ceilings interfere with maintenance of electrical and mechanical systems. Care must be taken to obtain the proper installation in areas where a fire resistive rating is required. In classrooms or meeting rooms, acoustic (sound) and illumination (lighting) considerations should be a priority.

Floor Finishes. Interior floors should be selected to reduce the likelihood of slipping and falling injuries, be easily cleaned, and durable. Vinyl tile, ceramic tile or carpet are appropriate. Choices of these material should be in heavy-duty grades. High-gloss concrete finishes should be avoided. Apparatus bay floors necessitate rugged yet slip resistant surfaces. Terrazzo (stone chips) is a very durable but expensive option. Seamless floors, usually epoxy with a sand aggregate, provide a durable floor, are easier to clean, and also minimize slipping hazards. Applied in multiple layers to a thickness of about one-eighth inch, they are wear resistant, easily cleaned, and slip resistant. If improperly installed, however, they may peel up. Sealed concrete can also be economical.

Heating, Ventilation, and Air Conditioning Systems. Heating, ventilating, and air conditioning (HVAC) systems make up a sizable portion of the construction, operation, and maintenance budget of fire stations. HVAC systems also impact occupant safety and health and should be selected carefully. Possible systems and relevant factors include:

- *Heating provided by electricity, fuel oil, natural gas, or propane fuel.* The availability of the fuel selected must be dependable. The fire hazards and potential leaks from storage of fuel oil or propane must be anticipated. Many codes now require the installation of costly containment and leak detection systems, for fuel storage tanks.

- *Heating provided via steam boiler systems.* Boiler systems provide a high-quality, comfortable, controllable, efficient and even heat for each type of use within the building, they are expensive to design and install.

- *Heating provided by hot water distribution systems.* Hot air systems are economical to install, especially when combined with the air conditioning duct work, but they can be drafty and subject to wide variations in temperature. They usually will not serve the apparatus bays well and must be supplemented.

- *Heating provide by direct fired, hot-air ducted or localized radiant heater systems.* Radiant heaters are well suited to high spaces such as apparatus bays, but inappropriate for the rest of the building. In colder climates the apparatus bay system must be capable of heating up quickly for returning frozen apparatus and personnel as quickly as possible.

- *Air conditioning using wall units or central systems.* Air conditioning through the wall units is only practical in very small stations. Central systems generally are more efficient and easier to control. In some areas of high electric rates, gas absorption units which burn gas to create cooling are being reintroduced with some success.

In addition, the following concerns exist for general station ventilation:

- Rapid removal of both exhaust fumes and combustible vapors in apparatus bays and prevention of fumes and vapors from reaching interior living areas or the compressor inlet for filling self-contained breathing apparatus. Approaches for achieving this removal include engineering controls on the apparatus, direct ventilation, and source capture methods.

- Ventilation of commercial grade kitchens via special hoods with grease filters, fire suppression systems, and make up air systems where called for by energy conservation requirements.

- Appropriate controls for obtain efficiency operation of HVAC systems which are not overly complex systems and are easily understood and maintained by department personnel.

Lighting and Electricity. Lighting on the apparatus floor is critical to prevent injuries and to complete work in a safe manner. The following provisions address the adequacy of lighting at the station:

- Apparatus floor lights should be controlled from the watch office as well as from the floor itself.

- There should be a provision for a single night light over each piece of apparatus which has its own separate control switch.

- Illuminating Engineering Society of North America (IESNA) standard LM 53-77 provides recommended illumination for interior spaces and classrooms,

- Since drills cannot always be held outdoors, fluorescent lighting should be provided for apparatus floors that will enable night drills to be held indoors.

- Sounding of alarms should be accompanied by actuation of lights. These lights should be separated by relevant area particularly for multi-purpose stations.

There are a number of general electrical needs which should be considered in the design of the fire/EMS station:

- Electrical outlets should be in ample supply.

- Duplex outlets should be installed with ground fault indicators.

- There should be outlets in all storage areas, locker rooms, and at least two outlets over work bench areas.

- Extra cabling or conduit should be allowed in the construction of stations to accommodate future needs for expanding electrical and cable capabilities.

Detection and Fire Suppression Systems. Given the role of most fire departments in promoting fire safety, it is not only sensible, but should be seen as a requirement for stations to be equipped with smoke and carbon monoxide detectors. Smoke detectors should be placed in all living spaces as well as work spaces. Carbon monoxide detectors should similarly be placed in living and work spaces. It is the position of the U. S. Fire Administration that all stations be equipped with automatic sprinkler systems.

Type of Construction

New Construction. In new construction, sites are typically identified with specific resource requirements. New construction generally represents an opportunity to properly design the station. However, budget constraints can restrict the implementation of special design features and improvements selected over past or existing stations.

Pre-Engineered Buildings. A municipality may have a set of pre engineered drawings if they are in a rapid expansion mode due to growth rate. Some cities have put trailers on sites with tent structures for the apparatus while stations were being built. Pre-engineered building for the most part are not the accepted practice for the fire service. The U.S. Army Corps of Engineers has developed a standard fire station design to provide guidance for one and two headquarters apparatus. While this concept seems to reduce design costs, special design considerations must be given to each specific site such as flooding conditions, seismic potentials and topography. There are nationally recognized consensus standards that should be built into each building such as NFPA's communications and life safety code standards as examples. With changing building codes, specialized needs, community cosmetics, multiple use structures, and varying locations, pre-engineered structures may require the same amount of design work to fit the location.

Remodelling. An area of primary concern is that stations should be built for potential growth. Cities can forecast their growth by looking at past years and have a good idea of where and how much their areas of responsibility will grow. This information may tell them a single engine or EMS facility is all that is needed or they may see a need to build a facility with more space than is needed right now. This also allows other stations the potential to use the extra space while their station is being remodeled or rebuilt. Consideration may be given to building stations with joint capabilities. This helps in deferring cost and allows the public better access to other community agencies.

SECTION 4 - SPECIFIC SAFETY AND HEALTH CONSIDERATIONS

There are several ways to examine a fire or EMS station in terms of safety and health. The conventional approach considers each space and the specific hazards related to that space. This approach lends itself to a design checklist or system of inspection. Each station can be divided into specific areas. For this manual, specific station areas are identified in Table 9.

Table 9. Station Building Areas*

A.   Station Grounds
  1.   Driveways
  2.   Training Areas and Training Pits
  3.   Vehicle Maintenance Areas

B.   General Station Interior
  1.   Public Areas
  2.   Office/Work/Watch Areas
  3.   Kitchen
  4.   Quarters/Toilets/Showers

C.   Laundry and Decontamination Areas

D.   Exercise/Gym Area

E.   Support Areas (Storage, Mechanical-Electrical Spaces, Shops)

F.   Apparatus Bays

G.   Special Areas
  1.   Mooring Docks (Fire Boat)
  2.   Hose Towers
  3.   Rappel Towers/Training

* Some areas do not apply to EMS stations.

A second approach to address station safety and health is to identify specific safety and health concerns. As would be expected, certain hazards are common to more than one area of the station. Thus, design solutions for one area may also be applied to other areas of the station. Of course, it is understood that some hazards are specific to certain station areas. One advantage of this approach is that it allows matching applicable regulations and code to specific hazards (Many regulations are non-area specific). In this manual, several applicable regulations have been identified which point to specific safety and health concerns (see Table 10).

Table 10. Safety and Health Concerns by Affected Station Area

Safety or Health Concern	Station Grounds	General Station Interior	Laundry/Decontamination Areas	Exercise/Gym Area	Support Areas	Apparatus Bays	Special Areas
Accidents							
Electrocution/Shock	■				■	■	
Slips/Falls	■	■			■	■	■
Explosions					■	■	
Vehicle	■					■	
Falling Objects				■	■	■	
Drowning							■
Overexertion				■			
Exposure Hazards							
Hazardous Materials			■		■	■	
Diesel Exhaust		■			■	■	
Smoke (Cigarette)	■						
Sick Building Syndrome		■	■	■	■		■
Radon Exposure		■					
Infectious Mat'ls/ Biohazards			■				
Food/Waterborne Infections		■					
Noise Pollution	■	■				■	
Other Hazards							
Natural Disasters	■	■	■	■	■	■	■
Fires (all types)	■	■	■	■	■	■	■
Criminal Activity							
Theft/Burglary	■	■	■	■	■	■	■
Vandalism/Violence	■	■	■	■	■	■	■

Electrocution/Shock Hazards

Nature of the Hazard. The possibility of electrocution exists wherever an electrical current contacts a person, usually through media such as water, wire, or other conductive mediums. Pathways for electric current may be confined to the limbs that contact the live circuit or more critically, the current may pass through the body, as in the case of hand-to-hand or hand-to-foot contact. This pathway has the most critical effect on heart function About 10 percent of the current from a hand-to-foot pathway flows through the heart. Since many processes within the body are mediated and controlled by electrical activity, external voltage can affect individuals resulting in a number of conditions. Threshold currents which cause these conditions range from 0.35 milliamperes (mA) generated by touching an improperly grounded appliance usually resulting in a tingling sensation to 100 mA from contact with a live electric power line which may result in head convulsions.

Extent of Problem at the Station. Several possible sources of electrocution arise in fire stations and older buildings:

1.   Wiring may be exposed, not being encased in conduit, being chewed by animals, exposed or otherwise frayed through age;

2.   Waterproof covers may be missing or outlets not at a proper height above ground;

3.   Electric power tools and outlets become ungrounded or are improperly used;

4.   Power lines in or around the station may be damaged or fall to the ground and come into contact with station personnel;

5.   Water contained in tankers and other fire fighting apparatus may leak onto the floor when tank seals age; any electrical device that comes into contact with this water can electrocute/shock an individual in and around the apparatus bays;

6.   Firefighters extending electrical lines without following codes; and

7.   Older fire/EMS stations not being able to meet today's increased electrical demands.

The most common source of electrical hazards in the fire station is maintenance once the systems are installed. Proximity of above-ground utility lines constitute electrical hazards when close to the station aerial ladders or towers. A number of firefighters have been seriously injured, one fatally, as the result of contacting an overhead power line at the station with an aluminum ground ladder in the past several years.

Relevant Regulations and Standards. The following are some the regulations and standards that apply to this area:

●   locally adopted building and fire codes
●   29 CFR 1910 137 Electrical protective devices

- 29 CFR 19 10.303 General requirements for electrical systems
- 29 CFR 1910.304 Wire design and protection
- 29 CFR 1910.305 Wiring methods, components, and equipment for general use
- NFPA *70, National Electrical Code*
- NFPA 70E, *Electrical Safety Requirements for Employee Workplaces*

Pertinent OSHA Regulations are provided in Appendix A under "All Areas - Electrical," including pages A-10 through A-14. Appendix B provides a list of federal offices in OSHA states while Appendix C provides the offices where state regulations can be obtained if in a non-OSHA state.

Specific organizations, listed in Appendix D, which provide guidance in this area include:

AEIC	Association of Edison Illuminating Companies
AHAM	Association of Home Appliances Manufacturers
AIA	American Institute of Architects
ALA	American Lighting Association
ASSE	American Society of Safety Engineers
EGSA	Electrical Generating Systems Association
ESD	Electrical Overstress/Electrostatic Discharge Association, Inc.
IEEE	Institute of Electrical & Electronics Engineers
LPI	Lightning Protection Institute
NECA	National Electrical Contractors Association
NEMA	National Electrical Manufacturers Association
NETA	International Electrical Testing Association
UL	Underwriters' Laboratories

Preventative Design Requirements. Electric shock injuries can be reduced by protecting personnel, selecting and maintaining appropriate equipment, and most importantly for this manual, designing the station or equipment to minimize the possibility of contact with electricity. All electrical work at the station must be done by a licensed electrician and then inspected by a qualified building/electrical inspector. Specific station design requirements include:

All Areas

1. Ground all station electrical outlets and connect to an electrical panel with circuit breakers, sized to handle the load on that circuit. Make sure that circuit breakers are clearly identified on the panel by function or area of coverage with a simple list on the panel door or a numbered facility diagram on the wall next to the panel.

2. Keep the main electrical panel area clear and free of storage (allow at least 3 feet of clearance). As with any facility, openings in the panel cover should be properly enclosed to avoid live parts from being exposed.

3. Ensure that all receptacles and junction boxes are covered with appropriate plates.

40

4. Install ground fault interrupt circuits-GFI (designed to interrupt the current during a short to reduce the time a person is in contact with it) in areas that are wet or could be wet, such as bathrooms, apparatus bays, boiler rooms, roofs, outside lighting circuits, and kitchens. Since receptacles can become damaged and wear out, periodically use a circuit tester, including a GFI tester, to ascertain their condition.

5. Do not rely on the frequent use of extension cords instead of fixed wiring. Add additional outlets as electrical needs are identified.

6. Protect light bulbs from physical damage when located within seven feet of the floor. Protection can be in the form of bulb sleeves or covers or plastic lenses.

7. Follow OSHA's lockout/tagout standard or NFPA 70E, *Standard for Electrical Safety Requirements for Employee Workplaces,* when personnel are working on energized circuits or deenergized fixed equipment.

Station Grounds

1. Ensure that utility poles (as obvious at this seems) are placed at least 10' to 15' away from vehicle maneuvering areas to reduce the possibility of collisions and the dropping of "hot" lines onto the vehicle or building. Large pumps located on the apparatus bumper may project up to three feet beyond the visible end of the vehicle. They can easily be missed when turning and come into contact with power poles.

2. Protect ground mounted transformers from impacts (Uniform Electrical Code) when placed within ten feet of driveways, or parking lots. A one hour fire resistive wall must be built to protect buildings or openings in them when occupants could be electrocuted, or burned by a damaged or failed transformer.

Support Areas

1. Ensure that all electrical powered tools or equipment are insulated to prevent an electrical shock to the user. Most portable or fixed equipment comes with a ground prong (pin) on the male attachment plug to provide a continuous path to ground if hazardous voltage escapes from the machine or equipment. If this prong is missing, the individual operating the tool or equipment may become the path to ground and receive a fatal electric shock. All electrical hand tools either must have a 3-prong plug (ground) or be labelled as "double insulated." Worn or missing plugs should be replaced with UL-listed plugs compatible with the cord and specific equipment.

2. Do not use 3-prong to 2-prong adapters ("cheater" plugs) to operate any equipment (cheater plugs are designed for allowing three prong plugs to fit into two prong outlets). NOTE: This requirement is not necessary if all station outlets are properly ground as required in item 2 above.

3. Design all block heater coaxial cords to feed from an overhead location, with a reinforced flexible connection to a junction box. Floor mounted cords can cause electrocution or become a tripping hazard.

4. Install an alarm-activated service disconnect of all fixed cooking devices for all new stations containing a kitchen and remodeled station kitchens. These devices shut off appliances when station personnel are mobilized by the alarm.

Slips and Falls

Nature of the Hazard. Personnel slips and falls occur when individuals loose balance or traction due to surfaces which are wet, uneven, or have poor traction, footwear that does not provide adequate slip resistance, or heights which unguarded. A variety of injuries can occur from these accidents. While most slips and falls result in strains, sprains, and broken bones, many can lead to debilitating injuries such as chronic lower back injury or even death.

Extent of Problem at the Station. Falls within fire/EMS stations or on station grounds account for a significant number of reported injuries within the fire service (excluding incident falls). Most occur during a response or training activities where rapid movements are required. Falls occur when a change of direction is required or where collisions with other personnel is possible, due to poor hallway layout, or where blind spots exist in the response line. Many areas of the station are prone to falling and slipping hazards such as:

*   Standing liquid on apparatus bay floors,
*   Limited maneuvering space between apparatus,
*   Crowded corridors and exits during response,
*   High step and improper rail design for some apparatus,
*   Poor tread design and lack of guard rails for stairways,
*   Lifting wet hose for suspension in the hose tower, and
*   Working around crowded docks or piers in marine settings.

Figures 5-8 illustrate a number of these problems at existing fire/EMS stations.

Elevated floor areas that have open sides require guard rails when the fall distance is greater than 6 feet (based on OSHA 29 CFR 1926.501). The probability of a fall resulting in traumatic injuries increases as the height increases and when the area below is a nonflexible surface such as hard-packed earth, concrete, or asphalt. The fatality rate increases dramatically at a lo-foot fall distance, according to OSHA statistics. Areas commonly in need of guard rails are hose tower platforms, storage areas, stairwell landings, outside areas such as decks, and possibly building roofs if the area is often used for training (including training towers). Guard rails form a system designed to prevent accidental falls.

A vertical hose tower hangs wet hose, folded in half, a distance of approximately 25 feet. The height of the hose tower generally allows the ends of handing hose to be at head level on the bottom of the tower. As a result, the height of the hose tower is 31 feet or more and the steel ladder going up the inside of the tower to provide access to the hose is often over 25 feet long.

Figure 5. Inaccessibility creates problems for responding fire fighters as shown. Between the tanker and fire truck lies a central column which is unprotected. Should a vehicle hit this, the upper structure of this station could easily collapse.

Figure 6. Evidence of water leaking across the accessible surfaces Within apparatus bay is a cause for concern for responding fire fighters to slip on wet surfaces

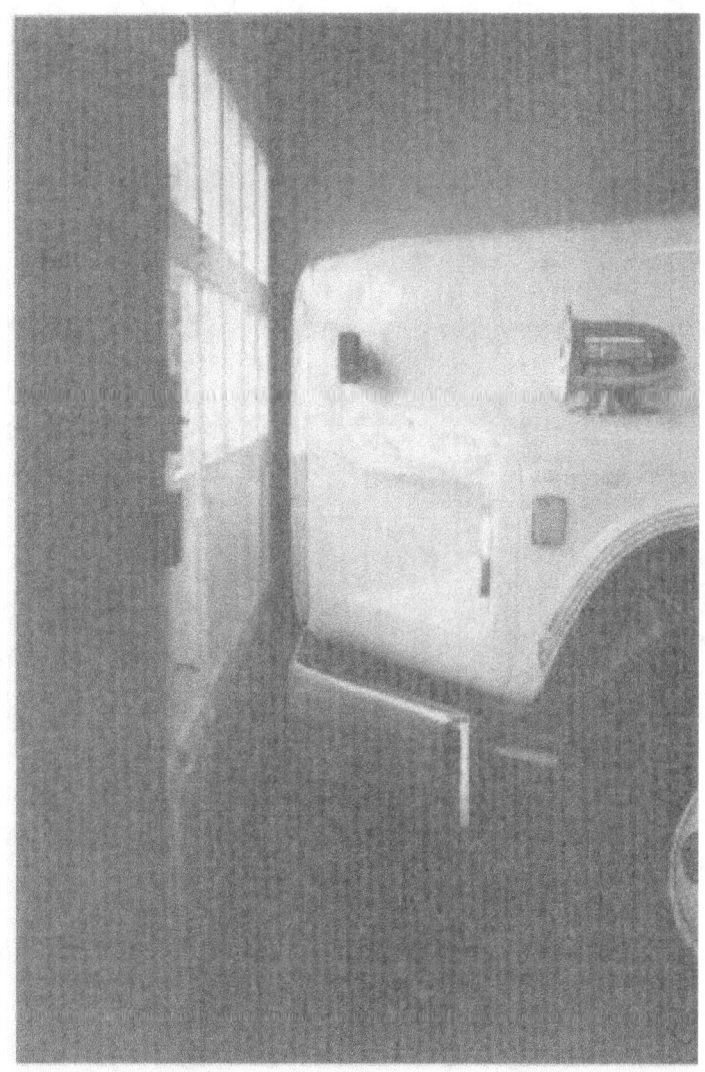

Figure 7. When adequate clearance is not provided, accessibility is limited and the likelihood of accidents is higher. This fire station was designed to meet vehicle requirements in the 1940s and is unable to adequately handle current fire apparatus dimensions.

Figure 8. The stairway leading to the second floor is routed down the center of the vehicle bay between the ambulance and one of the fire apparatus. Notice the stowed garbage can, emergency cones, and other janitorial items that are in the path of responding fire fighters.

Both OSHA and the American National Standards Institute (ANSI) have detailed specifications for straight ladders. These specifications generally are not followed when hose towers are built and not upgraded as technology evolves. OSHA and ANSI standards prohibit straight ladders in excess of 20 feet unless additional precautions are taken to reduce the hazard of falling off the ladder.

Relevant Regulations and Standards. The following are some the regulations and standards that apply to this area:

- locally adopted building and fire codes
- OSHA 1910.22 Walking-working surfaces, general requirements
- OSHA 1910.23 Guarding floor and wall openings and holes
- OSHA 1910.24 Fixed industrial stairs
- OSHA 1910.27 Fixed ladders
- OSHA 1910.36 Means of egress, general requirements
- OSHA 1910.37 Means of egress, general
- OSHA 1910.66 Power platforms for exterior building maintenance
- OSHA 1926.500-503, Subpart M, Fall Protection
- ANSI/ASSE *359.1-92, Safety Requirements for Personal Fall Arrest Systems*

Pertinent OSHA Regulations are provided in Appendix A under "Interior Areas - Walking-Working Surfaces" (pp. A-21 to A-31). Appendix B provides a list of federal offices in OSHA states while Appendix C provides the offices where state regulations can be obtained if in a non-OSHA state.

Specific organizations, listed in Appendix D, which provide guidance in this area include:

ACPA	American Concrete Pavement Association
AC1	American Concrete Institute
AEMA	Asphalt Emulsion Manufacturers Association
ANSI	American National Standards Institute
ASSE	American Society of Safety Engineers
AWPA	American Wood-Preservers' Association
HFES	Human Factors and Ergonomics Society
NTMA	National Terrazzo and Mosaic Association

Preventative Design Requirements. The majority of design solutions for preventing slips and falls involved providing appropriate walking surfaces with good traction, controlling interior station traffic, and providing guardrails and other fall restraint devices. Specific station design requirements include:

Station Grounds

1. Include in the design of circulation pathways, training areas and vehicle aprons, level, well drained, non-slip surfaces, constructed of either asphalt, concrete, corrugated steel plate, or fluid-applied membrane surfaces. (NOTE: Asphalt may not be appropriate for many heavy apparatus)

2.   Apply a non-slip texture to concrete using a light to heavy broom finish immediately after placement.

3.   Avoid painting large areas of asphalt surfaces since these surfaces are naturally non-slip except when painted or striped, or coated with oil droppings.

4.   Use a checkered plate or fluid applied surface containing sand or garnet for steel plates, often used for covering underground vaults, test pits, or transition surfaces, such as ramps.

5.   Require a coating of fluid applied deck coating containing sand or garnet for decks or other surfaces, including drill towers, which are rendered slippery when wet. Heavy traffic areas may require repeated fluid coatings as part of an ongoing maintenance program.

6.   Use run-off basins in areas for washing apparatus that are equipped with oil-water separators.

Station Interiors

1.   Require that access from the crew quarters to the apparatus bays be in a generally straight line with one or more access points provided.

2.   Avoid access hallways with turns or hallways that collect in a "T" or "X" crossing condition.

3.   Place access points in the apparatus bays which are near the front or rear of the bays. Place bunker gear lockers/storage at right angles to these access points to allow sufficient room for responding personnel to dress and not impede others.

4.   If possible, locate support and storage areas on the same level as the living and working spaces. The logic of this is often overlooked in both new or older stations. Support or storage areas that require the use of stairs to access mezzanine or loft areas, should be avoided. Falls occur when occupants carry items to be stored during lifting operations where unprotected openings are used. Vision is restricted on stairways or on "ladders" while grasping onto storage items.

5.   Three feet of clearance shall be maintained (designed) around apparatus parked within the station, if permitted by the structure. Aisles need to be established for spacing between apparatus much like the fire code requires aisle spaces in public assembly buildings, access to electrical panels, or aisles around high piled stock.

6.   Use slip-resistant surfaces on apparatus bay floors where personnel would normally mount or dismount apparatus (see Table 11).

Table 11. Notes on Floor Finishes for Prevention of Falls

Fire apparatus bay areas are finished with a smooth concrete surface and then given a seal coat. Serious consideration should be given to the need for non-slip surfaces which reduce the potential for slip-trip injuries in those areas which are frequently wet, and where firefighters are moving quickly, mounting, dismounting apparatus. Several manufacturers provide floor finishes that allow for safe, sure footing on wet surfaces. Once installed, the finishes remain permanently in place for extended periods and may be warranted for extended periods when properly maintained.

The installation of these floor finishes depends upon the manufacturer. Installation consists of applying special formulation of non-toxic chemicals which react with surfaces in such a way as to make porous mineral surfaces safe when wet. These treatments can be applied simply and safely to many types of existing floors both indoors and outdoors. The surface must be porous stone, i.e. granite, ceramic tile, quartz tile, terrazzo, concrete, brick, pool decking, marble, etc. These floor finish systems will not work on linoleum, vinyl tile or fiber glass. Some finishes may not be effective in greasy areas. They create a chemical reaction within the pores of the surface resulting in a higher coefficient of friction when water is present. One manufacturer's system cannot be detected on most surfaces after being applied. Another manufacturer's system uses an epoxy and quartz sand formulation that produces an non-skid or "orange peel" finish.

The application cost depends on the size of the area to be treated and the preparation time needed, and the local area (e.g., in Washington state, these costs generally run between $1.25/sq. ft. and $1.85/sq. ft. Cleaning and maintenance schedules will be different with floors treated with non-slip finishes.

One manufacturer offers a 5 year warranty and requires a regular maintenance program of wet mopping the treated surface with a manufacturers cleaning and reinforcing agent, in place of station normal cleaning solutions (annual maintenance cost ranges between approximately 2 cents to 55 cents per sq. ft. per year depending on the amount of soilage present and the number of times the surface is cleaned each week).

Several non-slip formulations will bring areas with slippery wet surfaces into compliance with safety standards set by OSHA, and into compliance with the Americans With Disabilities Act (ADA).

Stairways and Guardrails

1. Design guardrails with top rails at 42 inches, mid-rails at 21 inches, support posts at specified distances, and toe boards as needed. Guardrail construction should conform to OSHA and ANSI specifications, which allow latitude in construction materials. **(NOTE: These recommendations are based on OSHA requirements; the requirements in your area may be different; check your local building ordinances).**

2. Use non-slip finishes or treads on stair treads and the nosing (edge of the tread surface). Treads wear out and must be replaced as often as necessary to maintain a slip-resistant surface.

3. Use a color or hue on the edge of the stair tread (non-skid material) that contrasts with the rest of the tread.

4. Place guardrails on the open sides of stairways (to prevent falls over the side) when there are four or more risers. Stairway rail height should be between 30 and 34 inches measured vertically from the upper surface of the top rail to the surface of the tread in line with the face of the riser at the forward edge of the tread. The number of rails depends on the width of the stairs and the number of open or closed sides. Hand rails must be easy to grasp, no more than two inches in diameter, and at a height of not less than 34 and not more than 38 inches above the nosing of treads and landings. They must be continuous and extend beyond the last step. **(NOTE: These recommendations are based on OSHA requirements; the requirements in your area may be different; check your local building ordinances).**

5. Use contrasting colors for guardrails to make them clearly visible compared to the rest of the stairway.

Sliding Poles

1. Where sliding poles are used, guard the pole hole in such a manner as to prevent personnel from walking directly into the hole opening.

2. To absorb shock to sliding personnel, use a three foot diameter cushioned rubber mat or equivalent device at the bottom of all slide poles.

3. Consider using automatic door covers to remain in place over the sliding pole hole when not in use.

4. Establish rules which limit one person on the sliding pole at a time.

Elevated Surfaces (e.g., hose towers)

1. Equip floor openings in hose tower platforms with a forty-two inch guardrail with a mid-rail capable of withstanding a force of 250 lbs. applied in any direction at any point on the top rail. Equip each platform with toeboards (barriers at platform level erected along the exposed sides and ends to prevent materials falling off the platform).

2. When access ladders extend beyond twenty feet, require offset platforms and cage guards. (Permanent fixed ladders on the outside of drill towers and drill buildings are exempt from the requirements of offset platform landings and ladder cage guards).

3. Provide a rung clearance behind the hose tower ladders of at least seven inches.

4. Do not have step-across distances in excess of 12 inches.

5. Design side rails greater than 16 inches wide.

6   Use ladder rung spacings less than 12 inches.

7. Start ladder cages between seven and eight feet above a landing.

8. Use ladder pitch between 75 and 90 degrees.

9. Consider design alternatives for correcting a non-compliant hose ladder, including:
    *   installing a ladder safety device,
    *   installing alternate tread stairs to replace the ladder.
    *   installing a ladder cage,
    *   abandoning the use of the tower and using a hose dryer instead, or
    *   using a stairway access if one is already installed.

Marine Areas (Stations adjacent to bodies of water often use finger piers or docks to access water rescue craft.)

1. Construct the dock or pier surface using non-slip materials. Piers of concrete (chambered construction) require either a broom finish concrete or fluid applied non-slip coating.

2. Locate all tie-off points, hose bibs or electrical outlets away from walkways and provide adequate space for access to watercraft.

3. In locations where water levels fluctuate, require gangways to be designed for multiple angles of use. They must also be non-slip, using either raised treads of a non-slip surface or a combination of both. All gangways are required to meet (OSHA) handrail standards.

Explosions

Nature of the Hazard. Explosions occur when volatile vapors or gases come into contact with an ignition source or when pressurized cylinders rupture from overheating or physical breeches. Ignited gases send a front of flame and pressure in the area where gases have collected. Ruptured cylinders represent a physical or projectile like hazards. Both types of explosions are capable of causing severe personnel injury in the vicinity of the incident. Gas explosions can also ignite combustibles in adjacent areas, creating additional hazards.

Extent of Problem at Station. Refueling pumps and battery charging rooms pose a high risk of explosion or fire. Most common ignition sources are ungrounded outlets located less than 18" above the floor where vapor or gas accumulation occurs, or where equipment using unsealed switches is used in a room where gases have accumulated. An example of a potential problem is shown in Figure 9.

Other sources of explosions at the station involve pressurized cylinders or other vessels such as breathing apparatus bottles, fire extinguisher, and steam boilers. In 1993 one of the firefighters who died in the line of duty was killed by an exploding high-pressure air cylinder during a refilling operation. In 1986, a firefighter died in a steam boiler explosion in a fire station. In 1975, a firefighter was killed when a pressurized water extinguisher exploded.

Relevant Regulations and Standards. The following are some of the regulations and standards that apply to this area:

- locally adopted building and fire codes
- OSHA 1910.94 Ventilation
- OSHA 1910.101 Compressed gases (general requirements)
- OSHA 1910.106 Flammable and combustible liquids
- OSHA 1910.110 Storage and handling of liquified petroleum gases
- OSHA 1910.169 Air receivers
- NFPA *30, Flammable Liquids Code*
- NFPA *30A, Automotive and Marine Service Station Code*
- NFPA *54, National Fuel Gas Code*
- NFPA *55, Compressed and Liquified Gases in Portable Cylinders*
- NFPA *69, Explosion Prevention Systems*
- NFPA *85C, Furnace Explosions/Implosions in Multiple Burner Boiler Furnaces*

Pertinent OSHA Regulations are provided in Appendix A under "Compressed Gases" (p. A-33), "Hazardous Materials" (p. A-33 to A-35), "Apparatus Area" (p. A-35), and "Refueling Areas" (pps. A-37 to A-38). Appendix B provides a list of federal offices in OSHA states while Appendix C provides the offices where state regulations can be obtained if in a non-OSHA state.

Specific organizations, listed in Appendix D, which provide guidance in this area include:

AABC   Associated Air Balance Association
ABMA   American Boiler Manufacturers Association

Figure 9. This photo shows the battery charging area of a fire station, It was pointed out that the apparatus bay door was frequently left open to ventilate this area. However, during winter months, when batteries are charged more often, this is often not done and contributes to the possibility of a hydrogen build up.

AGA	American Gas Association
API	American Petroleum Institute
ASSE	American Society of Safety Engineers
BCI	Battery Council International
CAGI	Compressed Air and Gas Institute
CGA	Compressed Gas Association
IME	Institute of Makers of Explosives
NPGA	National Propane Gas Association

Preventative Design Requirements. In occupancy types with atmospheres that are classified as hazardous, the National Electrical Code (NFPA 70) cites specific requirements and locations of electrical equipment in these spaces.

Code Requirements for Hazardous Atmospheres

1. In all spaces where the possibility of an explosion exists, require mechanical ventilation of that space in accordance with the Uniform Mechanical Code Sections. The apparatus bays, shop areas, mechanical rooms, vehicle wash equipment reclaim rooms and electrical panels rooms present the highest risk of explosions in buildings ten years or older where modern safety code requirements were not designed into the building.

2. Locate gas lines, or regulators serving a building should be carefully researched. Regulators and manifolds should be located at least twenty feet (horizontally and vertically) away from fresh air intakes serving air handling equipment where burners or electric coils are used.

3. Consider the installation of explosive gas monitors or sensors in the design of new or remodeled fire or EMS stations for hazardous classified spaces.

4. Do not use Class I or Class II flammable liquids for cleaning purposes to remove grease or dirt from apparatus.

5. Install eye-washes near battery charging areas.

6. Install refueling pumps in accordance with the provisions of the locally adopted building codes or NFPA 30A.

7. Dispense Class I liquids as required by the locally adopted building code or NFPA 30A.

8. Post "No Smoking - Stop Your Motor" signs in fueling areas.

9. Place a refueling pump shut-off switch a minimum of 50 linear feet away from the dispenser and clearly post a sign, "Fuel Pump Shut-Off". This requirement is based on what is considered a safe distance away from the potential hazard area).

<u>Pressurized Cylinders and Vessels</u>

1. Hydrostatically test pressurized cylinders at regular intervals. The frequency depends on the cylinder construction. Common cylinders requiring hydrostatic testing are stored pressure fire extinguisher, breathing air cylinders, cascade cylinders, oxygen cylinders, and fuel gas (acetylene) cylinders,

2. Require inspection and certification by a boiler inspector for pressure vessels such as boilers, some hot water heaters, and some air tanks.

3. Secure cylinders in storage or use to prevent overturning. Use chain and rope for this purpose.

Vehicle Hazards

Nature of Hazard. The nature of activities in and around fire stations involves rapid vehicle movements as well as low speed backing or turning actions while responding to or preparing for a response to an EMS call. Accidents that occur as a result of these vehicle movements often involve firefighters or visitors to stations unfamiliar with these activities. The vehicles large bulk and many blind spots on fire fighting or EMS vehicles, further enhances the potential for an accident.

Extent of Problem at the Station. A significant number of incidents and injuries are related to vehicular use around stations. A number of these accidents can be eliminated by using a drive-through feature. Most in-station vehicular accidents are the result of apparatus backing up. Most fire departments rank vehicular accidents as the leading cause of station personnel injuries. Another type of accident occur when the apparatus move out onto adjacent roadways without proper traffic control. Lastly, apparatus bay doors are struck by apparatus or close on top of the apparatus. In many cases, the potential for injury to personnel is small; however, repair of doors can be difficult.

Relevant Regulations and Standards. Pertinent OSHA Regulations are provided in Appendix A under "Apparatus Area" (p. A-35). Appendix B provides a list of federal offices in OSHA states while Appendix C provides the offices where state regulations can be obtained if in a non-OSHA state.

Specific organizations, listed in Appendix D, which provide guidance in this area include:

AASHTO American Association of State Highway and Transportation Officials
ACPA   American Concrete Pavement Association
ARTBA   American Road and Transportation Builders Association
ATA   American Trucking Association
FHWA   Federal Highway Administration
ITE   Institute of Transportation Engineers
NAGDM   National Association of Garage Door Manufacturers
PSTC   Pressure Sensitive Tape Council

Figure 10. This rural fire station is located adjacent to a county highway. The distance between the apparatus bay doors and the highway is approximately 55 feet, allowing very limited access in and out of the apparatus bays. The bays are two-vehicles deep so that the returning vehicles must use the highway as part of their turnaround back into the station.

Preventative Design Requirements.

<u>Site Layout</u>

1. Whenever permitted by the site, design the station with "one way" circulation drives (usually one bay deep) which are accessed in a drive-through manner. Stations requiring a vehicle to back into the bays, will raise the risk of an accident, even when assistance is rendered by other firefighters. Avoid all backing operations where only the vehicle operator is available.

2. Equip driveways that empty directly onto a public street from the fire station in high volume traffic areas with traffic control lights and vehicle stop bars painted on either side of the driveway. For extremely busy roadways, there should be a series of lights in either direction in order to get through traffic. Activation of these lights is through the on board electronic signal control system.

3.   Avoid driveways that require more than one 90 degree turn to enter a public street.

4.   Install self-closing doors. These doors should have pressure switches at the bottom so in the event they start to close when an apparatus is under them, they return to the open position and do not damage the apparatus or the door itself.

Falling Objects

Extent and Nature of Problem. Injuries resulting from objects falling and striking personnel at the station are not uncommon in stations with storage mezzanines, lofts or high storage shelving or cabinets, or openings with sliding poles. The problem can be classified into two primary areas of concern, objects dropped by others and unsecured objects falling or toppling onto station occupants.

Figure 11. The storage within this station's apparatus bay for miscellaneous bulk items may be hazardous; some are located on top of a series of cabinets with a rail but placed in such a position that if an individual climbing the ladder should slop, or an article should drop, a fall or injury here is quite possible.

Relevant Regulations and Standards. Pertinent OSHA Regulations are provided in Appendix A under "Interior Areas - Walking-Working Surfaces, General" (pps. A-21 to A-26). Appendix B provides a list of federal offices in OSHA states while Appendix C provides the offices where state regulations can be obtained if in a non-OSHA state.

Specific organizations, listed in Appendix D, which provide guidance in this area include:

AAMA American Architectural Manufacturers Association
AIA American Institute of Architects
NSC National Safety Congress

Preventative Design Requirements. Areas most often associated with falling object accidents are the exercise gym, support or storage areas, and the apparatus bays. Items stored in or on file cabinets, storage shelving systems, wall mounted or hook supported systems represent the highest frequency of reported injuries from falling objects.

Exercise Room

1. Require supervision for the use of exercise gym equipment or at least implement a buddy system workouts when free weights are used in conjunction with standing or bench press maneuvers. The most common injuries involve the users loss of control of the free weight and its impact on the head, chest, extremities or neck areas. Without rapid assistance from a fellow firefighter, the injured user may sustain additional injuries or worsen injuries already sustained attempting to remove the free weight without assistance.

2. Allow for adequate floor space for two or more exercise participants at one time and all workout areas should be visible to day rooms or shift offices to ensure that victims can be seen and assisted quickly.

Storage Areas

1. Require all vertical files with three or more drawers to be of an interlock variety that allows only one drawer at a time to be opened. Secure all file cabinets to walls with lag screws and clamps or steel angles.

2. For storage shelving or high rise storage systems, require that all shelving have a one half inch lip on the open edge of each shelf. Install one retaining wire six inches above each shelf, and secure each section of a storage system to the floor or brace back to a wall or ceiling to prevent unit from falling.

3. For wall mounted items, use a ½ to 1 inch lips for objects stored on a hook or ledger to prevent the item from being unintentionally dislodged.

4. Require installation of walls or guard rails, that are a minimum of 42 inches high, on all mezzanine storage areas to prevent objects or people from falling to the level below.

5.   Control access to mezzanine areas by using a sliding or rolling gate at least four feet wide.

6.   Isolate loft spaces, which are accessed through a hinged door or gate, with three restraining cables, chains or bars to prevent occupants from falling through unprotected openings.

7.   Include a standard guardrail for a loft or landing platform with a toe board, which is a vertical barrier, at floor level. Erect guardrails along exposed edges of a floor opening, wall opening, platform, runway or ramp to prevent falls of material.

8.   In the design phase of station planning, always allow for additional storage space beyond what is then currently anticipated. In some cases, due to budget constraints and the cost of a station being assessed on a per square foot basis, it might be necessary to creatively assign space which can ultimately be used for storage.

Drowning

Nature of the Hazard. Drowning is a hazard at marine fire stations or for firefighters operating around piers or waterways. Contributing factors to drowning can include cold temperature water which causes hypothermia and limits individual mobility and the wearing of heavy protective clothing and equipment.

Extent of Problem at the Station. Stations which have marine settings may have the potential for drowning. Unguarded pier and dock locations combined with slippery conditions from either wet or cold weather pose the highest risks for drowning.

Preventative Design Requirements. The primarily means for preventing drowning is to follow the recommendations for slips and falls (see section above) and require that station personnel operating in unguarded areas around docks and piers wear a U.S. Coast Guard approved Personal Flotation Device (PFD).

Overexertion

Nature of the Hazard. Overexertion manifests itself as strains and sprains. These types of injuries may become long term and result in lost time. The most common types of injuries for this type include ankle sprains, tendinitis, and lower back pain.

Extent of Problem at the Station. Sports injuries have always plagued the fire service in lost time injuries. This is sometime partly due to using improperly designed sport facilities in some stations where no facility previously existed or sports activity on uneven playing surfaces. Other forms of overexertion come from improper lifting techniques for heavy objects. Strains and sprains often account for the largest source of station activity-related injuries.

Preventative Design Requirements. Proper playing equipment and areas should be established or alternative forms of exercise encouraged. Specific requirements include:

1. Encourage use of proper lifting techniques (see *Work Practices Guide for Manual Lifting*, NIOSH, 1981)

2. Establish specific locations for exercising and exercise equipment. Use only equipment which is approved by the department. Ensure that all personnel know how to use exercise equipment correctly.

3. Discourage sport activities in which improper surfaces or conditions exist. For example, basketball playing on uneven or slippery surfaces common on fire station aprons should be prohibited. Fire and emergency medical service personnel should instead be encouraged to participate in activities where the proper equipment and conditions are provided.

Hazardous Materials

Nature of the Hazard. Different chemicals present a variety of health, flammability, and reactivity hazards. The specific hazards vary with the chemical. While many chemicals may produce severe effects upon exposure at high concentrations, these same chemicals may create chronic health problems such cancer through repeated, low levels of exposure or even one-time, high concentration exposures.

Fire stations in some cities are used as decontamination sites for city workers that have been exposed to toxic or hazardous material. In this case it is necessary to have a designated area outside of the structure with hot and cold water and moderate enclosure along with a drain so these individuals will not contaminate the entire facility. This concept could also apply in general for many stations.

Extent of Problem at the Station. A variety of chemicals and solvents may be used around the station for maintenance and other applications. In addition, several fuels and lubricants may be at the station creating the potential for exposure at unacceptable levels.

Relevant Regulations and Standards. The following are some the regulations and standards that apply to this area:

- locally adopted building and fire codes
- OSHA 1910.94 Ventilation
- OSHA 1910.101 Compressed gases (general requirements)
- OSHA 1910.106 Flammable and combustible liquids
- OSHA 1910.110 Storage and handling of liquified petroleum gases
- OSHA 1910.1000 Air contaminants
- OSHA 1910.1200 Hazard communication
- DOT 14 CFR 103 and 49 CFR 171-179
- NFPA *30, Flammable Liquids Code*
- NFPA *55, Compressed and Liquified Gases in Portable Cylinders*

Pertinent OSHA Regulations are provided in Appendix A under "Hazardous Materials" (pps. A-33 to A-35). Appendix B provides a list of federal offices in OSHA states while Appendix C provides the offices where state regulations can be obtained if in a non-OSHA state.

Specific organizations, listed in Appendix D, which provide guidance in this area include:

ACGIH   American Conference of Governmental Industrial Hygienists
ACS   American Chemical Society
AICHE   American Institute of Chemical Engineers
AIHA   American Industrial Hygiene Association
API   American Petroleum Institute
ASSE   American Society of Safety Engineers
ASTM   American Society for Testing and Materials
CGA   Compressed Gas Association
NFPA   National Fire Protection Association
NPGA   National Propane Gas Association
NSC   National Safety Congress

Preventative Design Requirements.

<u>All Areas</u>

1. Locate all chemicals in specific, appropriate storage areas. Store flammable chemicals in a flammable chemical storage cabinet. Do not repackage chemicals.

2. Develop a written hazard communication program for the station which includes a list of chemicals used at the station.

3. Ensure that all chemicals are properly labeled and placarded with the appropriate warnings.

4. Maintain Materials Safety Data Sheets (MSDSs) on each chemical within the station area. Post the location of the MSDSs on the station bulletin board. Maintain free access to this location. Ensure that all personnel are aware of this location.

<u>Apparatus Bays</u>

1. Have oil or gas spilled within the apparatus area run through oil/water separators prior to entering the storm water system via the sewer system.

2. When fuel oil is required to be stored on site, choose above ground storage tanks before underground tanks. If possible, both systems require secondary containment which will store any oil that leaks out of the primary tank. The installation of either above or below ground tanks should be in accordance with locally adopted building and fire codes and state EPA regulations.

Diesel and Vehicle Exhaust Hazards

Nature of the Hazard. Diesel engines, used in fire apparatus, produce a mixture of toxic particulates and gases as the result of the combustion process. The composition of this exhaust product depends on several factors such as the specific fuel used, temperature of the engine, condition of the engine, cleanliness of the air intake filter, among others. An analysis of general diesel engine exhaust has revealed a variety of extremely toxic substances at significant concentrations, including:

> *Oxides of Nitrogen;* any combustion in air will produce various nitrogen oxides. Short term exposures can cause respiratory tract irritation and infections, Long term exposures result in lung tissue damage and difficulty in breathing.

> *Carbon Monoxide;* this chemical is produced as a by-product of combustion. Exposure to high levels of carbon monoxide causes death by tying up the hemoglobin in blood and preventing oxygen intake by the body. Exposure to carbon dioxide at lower concentrations causes headaches, dizziness, weakness, and neurological problems.

- *Volatile Organic Compounds* (VOCs); these compounds are a class of carbon-based chemicals such as benzene, toluene, phenol, and chlorinated solvents. Many of these chemicals cause a variety of adverse health effects such as headaches, nausea, neurological disorders, respiratory irritation, and liver damage. Some VOCs are known or suspected carcinogens.

- *Polyaromatic Nuclear Aromatics* (PNAs); PNAs are a class of relatively large, complex chemicals principally formed during the combustion processes. In diesel exhaust, these chemicals often adhere to the soot particles. Most PNAs are documented carcinogens.

Much of the diesel exhaust is invisible including the smaller soot particles. This means that exposure cannot always be detected. Furthermore, diesel exhaust can penetrate into clothing, furniture and other items with which firefighters have routine contact, where it can be later released after the initial exposure or absorb into the firefighters' skin. Continued exposure to diesel fuel emissions has been linked to cancer and other serious health disorders.

Extent of Problem at the Station. Both the National Institute for Occupational Safety and Health (NIOSH) and the U. S. Occupational Safety and Health Administration (OSHA) have declared human exposure to diesel exhaust as a potential occupational carcinogenic (cancer-causing) hazard through toxicological studies. A 1985 study commissioned by IAFF involved the measurement of diesel exhaust emissions at selected fire stations in New York, Boston, and Los Angeles. This study indicated that the most significant source of firefighter exposure to diesel exhaust was from the exhaust remaining in the station after the engine start. Some variations in the study results were identified, based on differences in climate, station design, number of runs per tour, and whether the firefighters smoked or not. From these findings, the IAFF study concluded:

Figure 12. Turnout gear stowed in the rear of this station near the access stairway is often exposed to diesel fumes and dust from the District's most frequently used vehicle-the EMS unit.

"Even with the uncertainties in the reported studies, apparent prudent public health practices would require that steps be taken to limit firefighter exposure to diesel emissions."

Later studies substantiated this problem to even show significant diesel exhaust emissions at smaller, less busy fire stations. As early as 1986, the New Jersey Department of Health distributed a bulletin alerting fire departments within the state of this problem.

Relevant Regulations and Standards. The following are some the regulations and standards that apply to this area:

- locally adopted building and fire codes
- OSHA 1910.94 Ventilation
- OSHA 1910.1000 Air contaminants
- NFPA *1500, Fire Department Safety and Health Program*

Pertinent OSHA Regulations are provided in Appendix A under "Hazardous Materials" (pps. A-33 to A-35). Appendix B provides a list of federal offices in OSHA states while Appendix C provides the offices where state regulations can be obtained if in a non-OSHA state.

Specific organizations, listed in Appendix D, which provide guidance in this area include:

AABC	Associated Air Balance Council
ACGIH	American Conference of Governmental Industrial Hygienists
ACS	American Chemical Society
AIHA	American Industrial Hygiene Association
ARI	Air Conditioning and Refrigeration Institute
ASHRAE	American Society of Heating, Refrigeration, and Air Conditioning Engineers
ASSE	American Society of Safety Engineers
ASTM	American Society for Testing and Materials
DEMA	Diesel Engine Manufacturers Association
HVA	Air Movement and Control Association
ICAC	Institute of Clean Air Companies
NFPA	National Fire Protection Association
NSC	National Safety Congress

Preventative Design Requirements. Various methods have been suggested for reducing diesel exhaust emissions at fire stations. These possible solutions take three different forms:

1. *Engineering controls* involve methods which reduce the amount of toxic substances released by the diesel engine.

2. *Ventilation* increases the flow of clean air to affected areas by adding apparatus room exhaust fans and in some cases providing a "positive pressure" in the living and work areas.

3. ***Source Capture*** entails placing collection systems directly on the apparatus tail pipe and venting the emissions harmlessly into the atmosphere.

Research has indicated that engineering controls offer the best method to reduce diesel exhaust emissions since this approach eliminates much of the hazard before it is generated. This solution is based on the use of cleaner burning fuels, better fuel ignition, and improved particulate traps. In addition, diesel engines built after 1995 have to meet more stringent emission standards set by the U.S. Environmental Protection Agency. Nevertheless, these solutions are currently not available and will be very expensive for the fire service to implement. A new generation of diesel engines will still produce some exhaust containing hazardous chemicals. Furthermore, as these newer diesel engines remain in service, their effectiveness in reducing exhaust will diminish as the engine no longer operates at its optimum performance levels. Interim solutions such as after market special filters attached to the apparatus tail pipe may appear to clean exhaust by removing visible particulates, but still allow hazardous gases to pass through and remain within the fire station.

Ambient "general" ventilation is the next approach. Strategies include diluting diesel exhaust gases with fresh air and keeping contaminated air from entering living or work spaces within the station. In its simplest form, the apparatus bay doors are left open for several minutes after the diesel-powered apparatus has left, allowing fresh air to enter the apparatus room. Typically, this approach also uses large capacity fans to exhaust diesel emission, creating a negative pressure within the apparatus bay, thus allowing clean air to rush in. Additional ventilation control can be provided by the station's HVAC system to "over" pressure living or work areas. However, the HVAC system must be carefully designed to remove all diesel exhaust because of "dead" air spaces within the station. Proper design of a ventilation system usually requires extensive modifications to the station at a very high cost. It is best considered during the construction of a new station. Moreover, this approach does not keep exhaust from absorbing into clothing and equipment (if stored in or near the apparatus bay) or other textile/plastic materials within the station. Since this type of control merely serves to dilute the fumes, it may not be an adequate solution.

Exhaust source capture is considered the most reliable means to reduce or eliminate exposure of fire station occupants to diesel exhaust emissions (Figure 13 shows an example of an installed system). This solution consists of a collection tube attached to the apparatus tail pipe with a high powered fan used to draw the exhaust gases through the tube to be discharged to the outside atmosphere. The more sophisticated systems include:

- an *automatic disconnect nozzle* allowing vehicles to drive into and out of the fire station with the hoses still attached to the exhaust system (the hose disconnects from the vehicle and retracts into the building allowing automatic doors to close);

- *automatic activation* via an in-line pressure switch when the apparatus engine is started or when apparatus bay doors open (ensuring consistent use); and

- *timers* to run exhaust fans until all residual gases have been removed from the system (manual operation is also provided).

More and more fire stations today are being designed with vehicle exhaust extraction systems in order to directly remove toxic vehicle emissions from the space at the source. This eliminates the problem of carbon monoxide vehicle exhaust contaminating the interior of the apparatus bay. While there are several systems that perform this function, they are all similar with the exception of how they are connected to the vehicles.

A newer version of an effective engine exhaust system can be built using the principles of down draft and positive pressure ventilation. In the system illustrated in Figure 14, the fire station is ventilated through a dual purpose ventilation and floor water drainage duct formed in the floor under each apparatus parking area. The slot forms a ventilation duct which is connected to high volume fans which both draw air out through the floor duct, and also pressure the apparatus bay with fresh air. The area heating unit is shut off momentarily during the exhaust cycle, which is timed to start when the bay doors are opened. This system leaves the work area clear of exhaust hose keeping the work area potentially safer and relatively free of obstructions. It also eliminates human factors and the fear associated with connecting or disconnecting fittings which can hang up when apparatus drive out.

Another aspect of station design relative to diesel exhaust which should be addressed is storage of personal protective equipment (PPE). The following preventative design requirements are recommended:

1. Store PPE away from the apparatus bay in a well-ventilated room. Since PPE is often wet following a response, the use of drying rooms is needed.

2. Due to the degradation of material when contacted by UV light, store PPE in a 'space free from sunlight.

Cigarette Smoke

Nature of the Hazard. The hazards of cigarette smoking are well known as cigarette smoking has been linked to lung cancer and other health disorders. Some public health studies indicate the harm of exposure to second-hand cigarette smoke.

Extent of Problem at the Station. A number of firefighters smoke, requiring designated smoking rooms or areas to prevent discomfort of fellow personnel. Some local ordinances may prohibit smoking in public facilities including fire and EMS stations.

Relevant Regulations and Standards. Most regulations and standards which apply to smoking areas in public facilities are local ordinances.

Preventative Design Requirements. Establish separate smoking areas or require smoking in outside areas only away from areas containing fuel or other flammable hazards (i.e., the apparatus bay area or storage areas).

Figure 13. This view of a station shows a source capture vehicle exhaust recoil system. The system extends forward as the vehicle leaves the apparatus bay and disconnects at the point of exit out of the overhead doors. This has proven to be a very effective system in fighting diesel exhaust fumes.

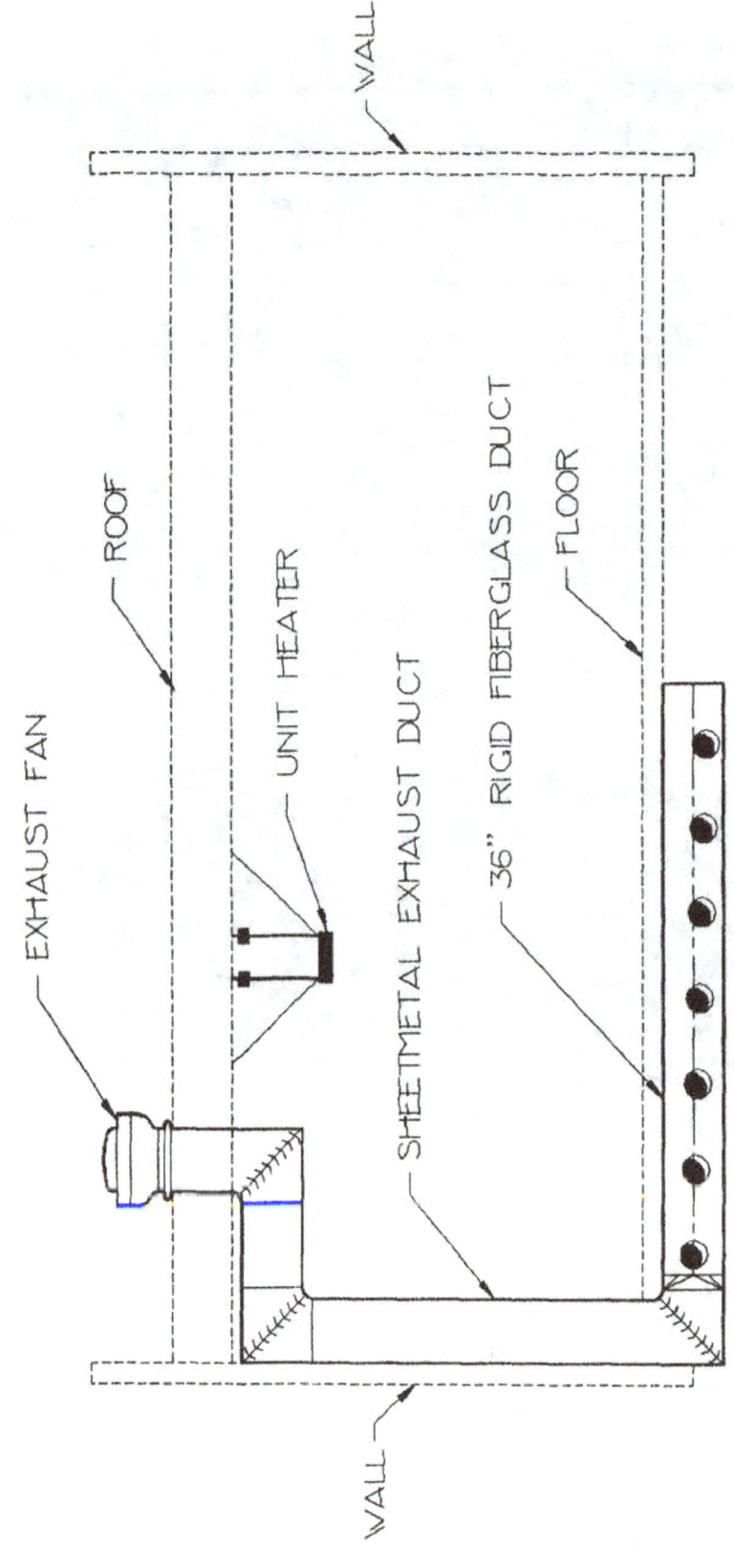

Figure 14. Design of In-floor Diesel Exhaust Source Capture System (courtesy of W. A. Barnard)

Sick Building Syndrome

Nature of the Hazard. Over the past several years, energy codes in colder regions of the country have become more restrictive in terms of higher insulation values and increased air tightness of new buildings. Illness and fatigue experienced by some inhabitants of these buildings (fire stations) can be traced to elevated levels of toxic gasses. These gasses can be emitted by building products, including adhesives, synthetic fabrics, paints, plastic materials, and many floor covering materials used in construction.

Extent of Problem at the Station. Offgassing or contamination within buildings can cause health problems, that relationship is well known. Studies have been run on three types of buildings-naturally ventilated, mechanically ventilated with operable windows and no air conditioning, and mechanically ventilated with air conditioning and sealed windows. Indoor and outdoor concentration of CO_2, CO, volatile organic compounds (VOC's), fungi and bacteria were measured in each along with indoor temperatures and humidifies,

VOC's give off high levels of vapors, so one is likely to inhale significant amounts of them. Many VOC 's are toxic. VOC's are extremely common. Any time materials are used in the construction process which are intended to dry out, there are almost always VOC's present. Paint, white out, paint stripper, almost any kind of cleaner fall into this category. Plastics are manufactured with VOC's, as are many other materials. It is probably not common for serious health problems to occur from exposure, except under unusual circumstances. Minor effects are very common.

Some research has shown that buildings using mechanical ventilation without air conditioning and those using mechanical ventilation with air conditioning had a higher prevalence of all symptoms except headaches as compared to building with natural ventilation. One possible explanation is that mechanical ventilation system are themselves sources of pollutants. The most common complaints concerning sick building syndrome are chills or fever, respiratory problems, dry or itchy skin, headaches, fatigue or sleepiness, and eye, nose, or throat irritation. There is a theory that the increase of carpet use adds to the sick building syndrome because of the multitude of micro biological agents that can nest in carpet fibers.

Fiberglass may also represent an area of concern. Fiberglass is simply strands of glass. The form of it known as glass wool, which is made of fine (small diameter) strands, is used as an insulating material and is the most commonly used insulating material in buildings, found in walls, ceilings, and ventilation systems. It is a part of a larger class of materials called man-made mineral fibers (MMMF) which include rock wool, made from fine strands of molten rock and slag respectively. Fiberglass inside ventilation systems should be sealed, or better yet replaced with a non fiberglass insulation. Fiberglass insulation in attics and walls should be covered in paper, foil, or best, plastic. Also pressure relationships should be maintained so air goes into the wall or attic, not out of it into occupied spaces. Probably the most major source of fiberglass contamination in buildings is the ventilation system. Nearly all ventilation systems have fiberglass exposed to air streams. It eventually breaks down and blows out.

Sick building syndrome describes the condition of a building in which more than 20% of the occupants are suffering from adverse health effects, but with no clinically diagnosable disease present. It is the condition of the building; not of the occupants. The 20% figure is arbitrarily set as there will always be some individuals complaining about adverse health effects associated with occupancy of a building. However, if the figure is 20% or more, it is considered that there must be some determinable cause which can be remedied.

The causes of SBS are still uncertain. Among the causes postulated, and for which some evidence exists, include man-made mineral fibers (MMMF), macromolecular organic chemicals (large size molecules produced by living things, such as protein molecules), volatile organic compounds (VOC's) and many others. Studies have shown that SBS is far more likely to occur in buildings which have air conditioning with an inadequate supply of fresh air than those whose fresh air supply satisfies current standards.

Some experts in the industrial hygiene community feel SBS is due to long-term exposure to low levels of a combination of contaminants. The level of each contaminant present may be far below any level at which a recognized health effect occurs, but the combination of numerous chemicals and particles at such levels than has the effect known as SBS. Research on this is at an early stage but tends to support this hypothesis.

Relevant Regulations and Standards. The Department of Labor, Occupation Safety and Health Administration published a proposed standard on Indoor Air Quality in April 1994 (29 CFR Parts 1910, 1915, 1926, and 1928 would be affected). Provisions in the standard are proposed to apply to all indoor work environments, requiring employers to develop a written indoor air quality compliance plan and implement that plan through actions such as inspection and maintenance of building systems that affect indoor air quality. The proposed standard covers control of specific contaminants such as hazardous chemicals, cleaning chemicals, microbial contamination, and pesticides. The standard will also require separate smoking areas which are enclosed and ventilate directly to the outside.

Specific organizations, listed in Appendix D, which provide guidance in this area include:

AABC	Associated Air Balance Council
ACGIH	American Conference of Governmental Industrial Hygienists
ACS	American Chemical Society
AICHE	American Institute of Chemical Engineers
AIHA	American Industrial Hygiene Association
AMCA	Air Movement & Control Association
ARI	Air Conditioning and Refrigeration Institute
ASHRAE	American Society of Heating, Refrigeration, and Air Conditioning Engineers
HVA	Air Movement and Control Association
ICAC	Institute of Clean Air Companies
I.E.S.	Institute of Environmental Sciences
NSC	National Safety Congress

Preventative Design Requirements. Architects and engineers play a significant roll in preventing or reducing the conditions contributing to illness or fatigue as caused by sick building syndrome. The architect can specify every material used in the construction of new or remodeling of older stations. Specifications should be written for materials and products that do not contain formaldehydes or toxic formulations. Mechanical Engineers are required to comply with national indoor air quality codes when calculating the amount of outside air required for ventilation inside buildings. Indoor air quality code revisions have been in effect since 1991, that require the use of 300 to 400 percent more outside air than was required prior to that time. These revisions help to balance the effect of "tighter buildings" on air quality with the need for efficient energy use.

1. At the completion of a new fire station or at the completion of remodel work and prior to occupancy, elevate the building's heating system setting to 100 degrees and maintain that temperature for a period of twenty-four hours while using low or no air changed to "cook" most gasses from newly installed products. This should be followed by a twelve hour period of high air changes to purge all spaces and system duct work.

2. Ensure that reviews made by fire district representatives during the design process address the newer standards which have been designed into their buildings. Currently the energy code requires the use of an economizer cycle on air handling units that supply over 3,400 cubic feet of air per minute. An economizer is used to bring in outside air for free cooling when temperatures are moderate. Economizers save substantial energy when compared to mechanical refrigeration processes.

Radon Exposure

Nature of the Hazard. Radon is a colorless, odorless, radioactive gas produced by the radioactive decay of radium-226, an element found in varying concentrations in many soils and bedrock. Because radon is a gas, it can easily move through small spaces between particles of soil and thus enter a building. Radon can enter a building as a component of soil gas and reach levels many times higher than outdoor levels. Radon levels are usually measured in picocuries per liter of air (pCi/L). Currently, it recommended that indoor radon levels be reduced to less than 4 pCi/L. But the lower the radon level, the lower the health risk; therefore, radon levels should be reduced to as closed to ambient levels as feasible (0.4 pCi/L). It can also accumulate in varying amounts in enclosed buildings. Radon is estimated to cause many thousands of lung cancer deaths each year. In fact, the Surgeon General has warned that radon is the second leading cause of lung cancer in the U.S. today

Extent of Problem at the Station. The most common way for radon to enter a building is from soil gas through pressure-driven transport. Radon can also enter a building through diffusion, well water, and construction materials (see Figure 15). The U.S. Environmental Protection Agency (EPA) has collected data on radon to compile a National Radon Potential Map. The map integrated five factors to product estimates of radon potential:

*   indoor radon screening measurements,
*   geology,
*   soil permeability,
*   aerial radioactivity, and
*   the type of substructure.

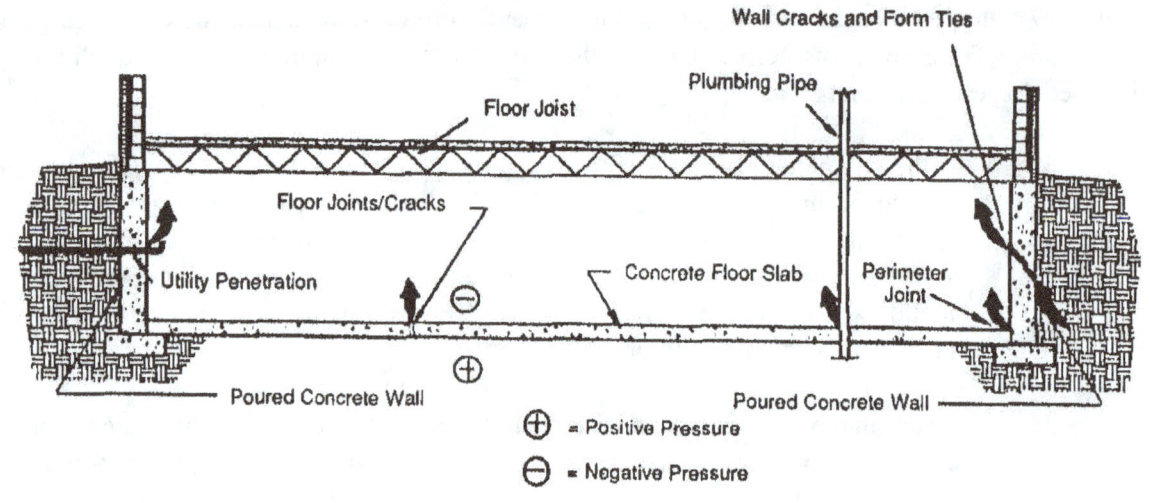

(a) Poured Concrete Basement Walls

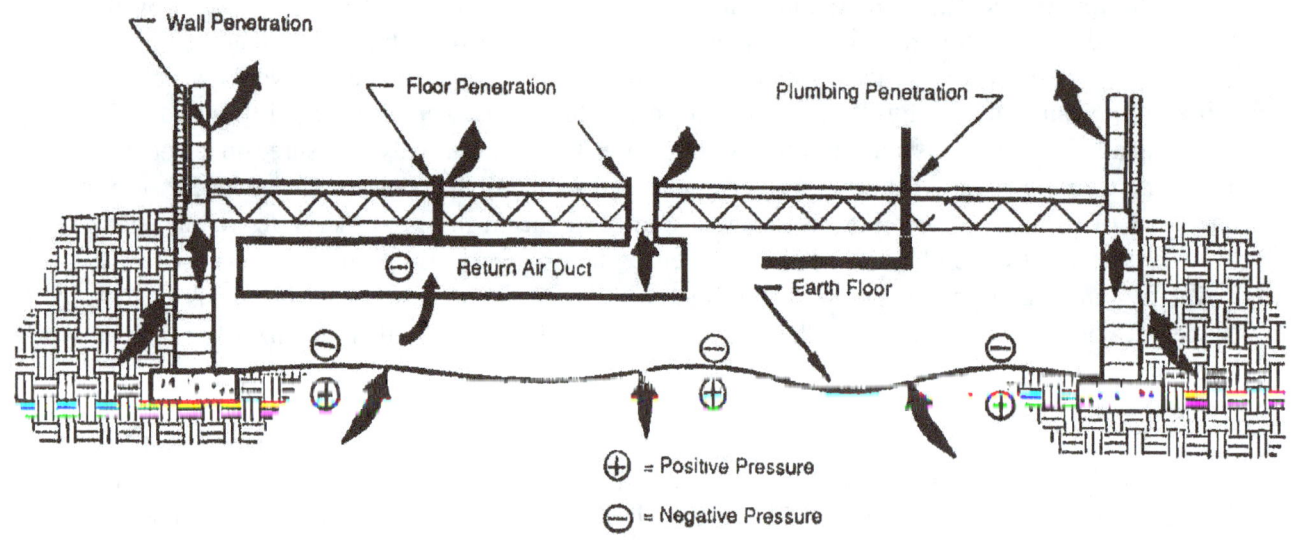

(b) Crawl Space Foundation

Figure 15. Typical Radon Entry Routes in Two Different Constructions

72

The EPA map assigns every county of the U.S. to one of three radon zones:

Zone 1 areas have the highest potential for elevated levels;

Zone 2 areas also have potential for elevated indoor radon levels but the occurrence is more variable; and

Zone 3 areas have the least potential for elevated levels.

The EPA has surveyed a sample of existing U.S. school buildings and has found many schools and other large buildings with radon levels above 4 pCi/L and some higher than 20 pCi/L. Several studies have attempted but have failed to make simple correlations between radon or radium concentrations in the soil and indoor radon concentrations.

Relevant Regulations and Standards. The EPA has established "The Indoor Radon Abatement Act of 1998." As stated in the act, the national long-term goal of the United States with respect to radon levels in buildings is that the air within the buildings should be as free of radon as the ambient air outside the building. Additional information on radon can be obtained from Radiation Program Manager or the specific EPA Regional Office (see Table 12 for a listing).

Specific organizations, listed in Appendix D, which provide guidance in this area include:

AABC   Associated Air Balance Council
ACGIH   American Conference of Governmental Industrial Hygienists
AIHA   American Industrial Hygiene Association
AMCA   Air Movement & Control Association
ASHRAE American Society of Heating, Refrigeration, and Air Conditioning Engineers
ASSE   American Society of Safety Engineers
ASTM   American Society for Testing and Materials
BIA   Brick Institute of America
HVA   Air Movement and Control Association
ICAC   Institute of Clean Air Companies
NSC   National Safety Congress

Preventative Design Requirements. Like most other indoor air contaminants, radon can best be controlled by keeping it out of the building in the first place, rather than removing it once it has entered. The three primary design/engineering means for achieving this include:

1. *Soil Depressurization.* A suction fan is used to produce a low-pressure field under the construction slab. This low-pressure field prevents radon entry by causing air to flow from the building into the soil.

2. *Building Pressurization.* Indoor/subslab pressure relationships are controlled to prevent radon entry. More outdoor air is supplied than exhausted so that the building is slightly pressurized compared to both the exterior of the building and the subslab area.

3.   ***Sealing Radon Entry Routes.*** The major routes for radon entry are blocked.

The most effective and frequently used radon reduction technique in existing building is active soil depressurization (ASD). An ASD system creates a low-pressure zone below the slab by using a powered fan to create negative pressure beneath the slab and foundation. This low-pressure field prevents soil gas from entering the building because it reverses the normal direction of airflow where the slab and foundation meet. If the low pressure zone is extended through the entire subslab area, air will flow from the building into the soil, effectively sealing slab and foundation cracks and holes. Several factors affect the coast of an active soil depressurization system. Incremental installation costs for a system designed into a new large building range from $0.10 per square foot of earth contact to more than $0.75 per square foot of earth contact. In comparison, the cost of installing an ASD system into an existing structure can range from $0.50 per square foot of earth contact to more than $3 .00 per square foot of earth contact depending on the structure and subslab materials.

Difficulties in providing radon protection in an existing structure can include:

- poor communication below the floor slab (i.e., no aggregate or aggregate with many fines or wide particle size distribution);

- barriers to subslab communication (internal subslab walls);

- radon entry points at expansion or control joints;

- ease of running the radon vent pipe and power source through and/or out onto the building's roof; and

- building depressurization caused by the HVAC system (or other fans) exhausting more air than is supplied.

Detailed guidelines for design and construction technique to minimize radon exposure are contained in EPA Publication, "Radon Prevention in the Design and Construction of Schools and Other Large Buildings," EPA/625/R-92/016, June 1994. Request in publication in writing from U.S. EPA, Center for Environmental Research Information, Cincinnati, OH 45268; provide title and publication number.

Table 12. EPA Regional Offices and Contacts for Radon Remediation

Region 1 (CT, ME, MA, NH, RI, VT) JFK Federal Building Boston, MA 02203 Attention: Radiation Program Manager 617-565-4502	**Region 6** (AR, LA, NM, OK,TX) 1445 Ross Avenue Dallas, TX 75202 Attention: Radiation Program Manager 214-655-7223
Region 2 **(NJ, NY)** 26 Federal Building New York, NY 10278 Attention: Radiation Program Manager 212-264-4418	**Region 7** (IA, KS, MO, NE) 726 Minnesota Avenue Kansas City, KS 66101 Attention: Radiation Program Manager 913-551-7020
Region 3 (DE, DC, MD, PA, VA, WV) 841 Chestnut Building Philadelphia, PA 19107 Attention: Radiation Program Manager 215-597-8320	**Region 8** (CO, MT, ND, SD, UT, WY) 999 18th St., Denver Place, Ste. 500 Denver, CO 80202 Attention: Radiation Program Manager 303-293-1709
Region 4 (AL, FL, GA, KY, MS, NC, SC, TN) 345 Courtland St., N.E. Atlanta, GA 30365 Attention: Radiation Program Manager 404-347-3907	**Region 9** (AZ, CA, HI, NV) 75 Hawthorne Street San Francisco, CA 94105 Attention: Radiation Program Manager 415-744-1045
Region 5 (IL, IN, MI, MN, OH, WI) 77 West Jackson Boulevard Chicago, IL 60604 Attention: Radiation Program Manager 800-621-8431 (except IL) 800-572-25 15 (IL)	**Region 10** (AK, ID, OR, WA) 1200 Sixth Avenue Seattle, WA 98101 Attention: Radiation Program Manager 206-442-7660

Infectious Materials and Biohazards

Nature of the Hazard. A number of microorganisms exist in food, water, and sewage. If unsanitary conditions exist, certain microorganisms can produce disease in humans (e.g., Hepatitis A Virus). Infectious agents in blood and other body fluids can be transferred by direct and indirect exposure. Bloodborne pathogens, such as Human Immunodeficiency Virus (HIV) and hepatitis, are transferred by direct contact with blood and certain other body fluids with mucous membranes and abraded or cut skin of the emergency responders. The International Association of Fire Fighters (IAFF) estimates that 1 out of every 15 firefighters has an exposure to a HIV-infected patient each year.

Extent of Problem at the Station. Firefighters and EMS personnel face exposure to HIV and hepatitis during emergency medical operations. While much of the risk of exposure occurs during the incident, the subsequent contamination of clothing and other equipment creates concerns for potential infection of personnel at the station, particularly during cleaning of these items. Departments are increasingly considering various methods for including appropriate laundry facilities at different stations.

Relevant Regulations and Standards. The following are some the regulations and standards that apply to this area:

n   locally adopted building and fire codes
n   OSHA 1190.1030 Bloodborne pathogens
n   NFPA 1581, *Fire Department Infection Control Program*

Pertinent OSHA Regulations are provided in Appendix A under "Medical-Related Areas" (pps. A-19 to A-21). Appendix B provides a list of federal offices in OSHA states while Appendix C provides the offices where state regulations can be obtained if in a non-OSHA state.

Specific organizations, listed in Appendix D, which provide guidance in this area include:

AAMI   Association for the Advancement of Medical Instrumentation
AIHA   American Industrial Hygiene Association
APHA   American Public Health Association
ASTM   American Society for Testing and Materials
NFPA   National Fire Protection Association
NSC   National Safety Congress

Preventative Design Requirements. NFPA 1581, *Standard on Department Infection Control Program,* specifies several requirements for the procedures within special facilities of the station for handling biocontaminated clothing and equipment.

Facilities For Disinfecting - Fire and EMS departments that provide emergency medical operations shall provide or have access to disinfecting facilities for the cleaning and disinfecting of emergency medical equipment. Medical equipment shall only be disinfected at a facility that meets the disinfection requirements of this section. Disinfection should not be conducted in station kitchen, living, sleeping or personal hygiene areas.

1. Design disinfecting facilities in stations with proper lighting, separate ventilation to the outside environment, fitted with floor drains connected to a sanitary sewer system, and to prevent contamination of other station areas.

2. Within disinfecting facilities, install a minimum of 2 sinks with hot and cold water faucets and a sprayer attachment, and with drains connected to a sanitary sewer system. Sink faucets should not require the user to grasp, with hands, to turn on or off. All surfaces should be nonporous material with continuous molded counter top and splash panel surfaces.

3. Equip disinfecting facilities with rack shelving or nonporous material. Shelving should be provided above sinks for drip-drying of cleaned equipment. All drainage from shelving should either go into a sink or drain directly into a sanitary sewer system

4. If possible, select front loading industrial laundry machines designed for the type of cleaning required for protective clothing.

5. When exposure occurs, clean the equipment and store the waste water from this process in a double wall tank where it can then be pumped to waste transfer vehicles for appropriate disposal.

Cleaning Areas

1. Provide a designated cleaning area in each station for the cleaning and disinfecting of protective clothing, protective equipment, portable equipment, and other clothing. This cleaning area should have proper ventilation, lighting, and drainage connected to a sanitary sewer system.

2. Physically separate the designated cleaning area from areas used for food preparation, cleaning of food and cooking utensils, personal hygiene, sleeping, and living areas; also physically separate the designated cleaning area from the emergency medical disinfecting facility.

Storage Areas

1. Store station emergency medical supplies/equipment, other than that stored on vehicles, in a dedicated, enclosed room protected from the outside environment.

2. Store protective clothing and protective equipment in a dedicated, well-ventilated area or room.

3. Do not store reusable emergency medical supplies and equipment, protective clothing, and protective equipment in a kitchen, living, sleeping or personal hygiene areas, nor shall it be stored in personal clothing lockers.

Food and Waterborne Infections

Nature of the Hazard. A number of microorganisms are formed in spoiling food and sewage. If unclean conditions persist, certain microorganisms can grow at rates sufficient enough to harm humans. Other unwanted microorganisms and other contamination can enter the station through cross contamination of water supplies. Cross connections are links through which it is possible for polluting or contaminating materials to enter a potable water supply. Contamination enters the potable water system when the pressure of the polluted source exceeds the pressure of the potable source.

Extent of Problem at the Station. Station personnel may be susceptible to food and waterborne based infections through improper food storage and back flow from sewage systems. At the fire or EMS station, cross contamination may occur in laundry facilities, sewage systems, and from outdoor operations involving training.

Relevant Regulations and Standards. The following are some the regulations and standards that apply to this area:

- locally adopted building and fire codes
- OSHA 1910.141 Sanitation
- NFPA 1581, *Fire Department Infection Control Program*

Pertinent OSHA Regulations are provided in Appendix A under "Sanitation" (pps. A-15 to A-17). Appendix B provides a list of federal offices in OSHA states while Appendix C provides the offices where state regulations can be obtained if in a non-OSHA state.

Specific organizations, listed in Appendix D, which provide guidance in this area include:

APHA	American Public Health Association
APWA	American Public Works Association
ASTM	American Society for Testing and Materials
AWWA	American Water Works Association
CMA	Cookware Manufacturers Association
CRMA	Commercial Refrigerator Manufacturers Association
KCMA	Kitchen Cabinet Makers Association
NSWMA	National Solid Waste Management Association
PDI	Plumbing and Drainage Institute
SSPMA	Sump and Sewage Pump Manufacturers Association
WQA	Water Quality Association

Preventative Design Requirements. This kitchen area of a fire or EMS station is often the most neglected. Traditionally individuals who held the purse strings thought spending anything for comfort or health was a waste of taxpayers money. Often ranges or refrigerators had to be replaced many times rather than buying what was needed in the first place. Besides being expensive to replace, improper kitchen equipment and design can contribute to unsanitary conditions.

Kitchen

1. Use ranges that are commercial grade, self cleaning and that include hoods.

2. Use commercial grade dishwashers.

3. Select refrigerators which are commercial grade and most often two are needed because firefighters work in rotating shifts. Each shift should have their own refrigerator.

4. Require double sinks which have heavy duty grade garbage disposals with matching rated drains.

5. Provide separate storage cabinets for food storage. Allow one storage cabinet for each shift.

6. Design the kitchen layout with a center island counter for the preparation of food and availability to food once it is served.

Sanitary Facilities

All sanitation requirements must meet state and local requirements. Each station should have gender specific facilities and a water closet for the public. Shower stalls should be enclosed on three sides to provide some degree of privacy. Place locks on station restroom doors to ensure privacy.

Public stations must have ADA-compliant toilet facilities with the appropriate fixtures and access.

Prevention of Cross Contamination

A backflow on water outlets directly connected to a hose can cause back pressure or back siphonage. The solution requires the elimination of the cross connection and the installation of an air cap or back flow-prevention assembly. Several methods can be used for the prevention of back flow, including:

- air gap,
- reduced-pressure principle back flow-prevention assembly,
- double check valve assembly,
- pressure vacuum breaker, and
- atmospheric vacuum breaker.

An air gap is the unobstructed vertical distance through air between the lowest point of a water supply outlet and the flood level rim of the fixture or assembly into which the outlet discharges.

The reduced-pressure principle back flow-prevention assembly consists of two independently acting, approved check-valves together with a hydraulically operating, mechanically independent pressure differential relief-valve located between the check-valves and below the first check-valve.

A double check-valve assembly consists of two internally loaded check-valves, either spring-loaded or internally weighted, installed as a unit between two tightly closing resilient-seated shutoff valves as a assembly.

A pressure vacuum breaker consists of an independently operating internally loaded check-valve and an independently operating loaded air inlet valve located on the discharge side of the check-valve.

An atmospheric vacuum breaker is an assembly that performs similarly to a pressure vacuum breaker. The atmospheric vacuum breaker consists of a float check, a check-seat, and an air inlet port. A shutoff valve immediately upstream may be an integral part of the assembly.

Detailed designs and practices for prevention of back flow and for cross-connection control are provided in the American Water Works Association Manual of Water Supply Practices AWWA M14, "Recommended Practice for Backflow Prevention and Cross-Connection Control." Obtain a copy of this publication by writing to the AWWA (see Appendix D).

Noise

Nature of the Hazard. Excessive noise can affect firefighters or EMS personnel in any of the following ways:

- It may contribute to hearing loss;
- It may interfere with communication;
- It may annoy or distract individuals; and
- It may alter performance on some tasks.

Criteria have been established by several sources for defining levels of acceptable noise in the workplace:

1. *Hearing loss* - Although it occurs gradually, hearing loss represents irreversible damage to the inner ear. The degree to which hearing is affected depends on the intensity, frequency spectrum, and duration of noise exposure, plus individual susceptibility. Noise-induced hearing loss usually produces loss of the high frequency components first, resulting in reduced quality, clarity, and fidelity of sounds. Research has indicated that extended exposure, about 8 hours, to noise levels in excess of 85 dBA (decibels on the A scale of a sound-level meter) may cause hearing loss. Most of the these studies have provided data characterizing noise exposures during emergency medical operations, some exceeded 120 dBa for short times.

2. *Annoyance and distraction* - Although noise levels below 85 dBA probably do not contribute to hearing loss problems, they may contribute to performance decrements due to distraction or annoyance. Noisy equipment can reduce the effectiveness of communications and make it difficult for people to concentrate in some types of tasks.

3. *Interference with communication* - Speech interference by noise is fairly common around noisy machinery or vehicles. The criticalness of communications should determine the steps to be taken to decrease the noise levels in the environment,

4. *Impact on work performance* - Noise may affect performance in a number of ways. Noise features which are most likely to degrade performance include variability in level or content, intermittency, high-level repeated noises, frequencies above 2000 Hz, or any combination of these features.

Several studies of noise exposures have been conducted for the fire and emergency medical services.

Noise measurements have generally been divided into four categories:

1. siren operations;
2. incident scene environment;
3. return travel to station; and
4. station environment.

Log book records were used to estimate times at the station and the number of emergency runs. The calculated noise levels suggest that emergency responders were exposed to noise levels that exceeded the OSHA Permissible Exposure Limit (PEL) of 90 dB(A). These excessive exposures were found for firefighters occupying nearly all of the riding positions on the vehicles.

Extent of Problem at the Station. While advances have been made in combating noise inside emergency apparatus, fewer improvements have been made inside the station. This is becoming more of a problem as stations use designs with hard surface floors, walls, and ceiling for durability, maintenance, and cleaning; these materials reflect rather than absorb noise from sirens or alarms, televisions, and radios. Even the living quarters of firefighters can be noisy, being located on or near major highways for easy access to roads. Many kitchens have metal stoves and heavy metal cooking utensils that contribute to the overall noise emissions. Departments that have the additional responsibility of airport fire and rescue also hear the aircraft landings and take-offs.

Relevant Regulations and Standards. The following are some the regulations and standards that apply to this area:

- locally adopted building and fire codes
- OSHA 1910.95 Occupational noise exposure
- NFPA 1500, *Fire Department Occupational Safety and Health Program*

Pertinent OSHA Regulations are provided in Appendix A under "Noise" (p. A-18). Appendix B provides a list of federal offices in OSHA states while Appendix C provides the offices where state regulations can be obtained if in a non-OSHA state.

Specific organizations, listed in Appendix D, which provide guidance in this area include:

ACGIH   American Conference of Governmental Industrial Hygienists
AIHA   American Industrial Hygiene Association
APHA   American Public Health Association
HFES   Human Factors and Ergonomics Society
NFPA   National Fire Protection Association
NSC   National Safety Congress

Preventative Design Requirements. Four different methods can be used for the reduction of noise:

1.   Alter the spectrum of generated noise. This approach can be used for the selection of internal station alarms. System alarms should be at an appropriate loudness levels and tied in with visual alarms to reduce dependence on the auditory alarm alone.

2.   Use barriers such as strategically placed walls or station areas to reduce noise transmission through the structure. Living quarters with the station should be buffered from day rooms and other areas where background noise is common. Airport stations or stations next to busy highways should include a noise barrier between the station and the source of noise.

3.   Locate noisy equipment away from living areas if permitted within the structure. Laundry facilities, pumps, and SCBA compressors should be separated in the layout of the station.

4.   Absorb incident or reflected noise by using acoustic ceiling tiles and carpet help inside the station.

Design of the station to minimize noise must also consider noise which emanates from the station to the community by using natural barriers when possible.

Federal Emergency Management Agency/U.S. Fire Administration Report FA-118 (November 1992), entitled "Fire & Emergency Service Hearing Conservation Program Manual" addresses different ways for minimizing noise at the fire/EMS station. Request a copy of this publication by writing to the Fire Management & Technical Programs Division, U. S. Fire Administration at 16825 South Seton Avenue, Emmitsburg, MD 21727.

Earthquakes and Other Natural Disasters

Nature of the Hazard. The effects of earthquakes on structures are well known. Current codes divide the country into four seismic zones, ranging from high risk to non-existent or low risk. Current codes do not deal with the variables of geology within a seismic zone. Amplifications, caused by locally soft soils, can create much of this variability. This variability within a seismic zone can vary significantly over a short distance. Good geological data is vital in developing designs for new fire stations or evaluating the survivability of an existing one.

Other natural disasters are possible throughout the country, including the effects of high winds, flooding, and lightning associated with severe storms. The likelihood of these events will differ by the region in the country as well as the specific location of the station. High winds pose particular problems for roofs and outside structures. Flooding will primarily affect low lying areas or communities next to rivers or lakes. Lightning and severe storms can occur anywhere but areas prone to tornados and hurricanes are especially at risk.

Fire and EMS stations are often thought of as a place of safety and shelter. Consideration should be given to one station in a city as the focal point for disaster should a catastrophic event happen. This may dictate the need for a special conference room where all local government officials meet and plan the survival of the city in the event of such a happening. This requires extra phone lines, rest rooms, chalkboards, break out rooms, and ample storage for emergency supplies.

Extent of Problem at the Station. Fire and EMS stations are considered by most building codes "essential structures" and therefore their design is vital for the safety of the emergency response personnel and the citizens who depend on them. Essential structures are designed to 125% of the earthquake resistance standards required by other residential and public structures. Those basic standards range from level one to level four, depending on the seismic zone risk in a particular area. In addition, survivability of the station must be considered for severe weather. For example, appropriate station locations must be considered in areas where flooding is possible.

Relevant Regulations and Standards. The primary source of regulations in the area of withstanding earthquakes and other natural disasters is the locally adopted building code. Specific organizations, listed in Appendix D, which provide guidance in this area include:

AC1	American Concrete Institute
AIA	American Institute of Architects
AISC	American Institute of Steel Construction
AISG	American Insurance Services Group
AITC	American Institute of Timber Construction
ASCE	American Society of Civil Engineers
CED	Civil Engineering Data
CRSI	Concrete Reinforcing Steel Institute
CSI	Construction Specifications Institute
FGCC	Federal Geodetic Control Committee
IRI	Industrial Risk Insurers

PC1   Precast/Prestressed Concrete Institute
PTI   Post Tensioning Institute

Preventative Design Requirements.

Earthquakes

In the design of a building to resist earthquake motions, the designer works within certain constraints, such as the architectural configuration of the building, the foundation conditions, the nature and extent of the hazard should a failure or collapse occur, the possibility of an earthquake, the possible intensity of earthquakes in the region, the cost or available capital for construction, and similar factors. The designer must have some basis for the selection of the strength and the proportions of the building and of the various support members in it. The required strength depends on factors such as the intensity of the earthquake motions to be expected, the flexibility of the structure. and the ductility or reserve strength before damage occurs. Because of the interrelations among flexibility and the strength of a structure, and the forces generated in it by earthquake motions, the dynamic design procedure must take those various factors into account. The ideal to be achieved is one involving flexibility and energy-absorbing capacity which will permit the earthquake displacements to take place without generating unduly large forces. To achieve this end, control of the construction procedures and appropriate inspection practices are necessary. The attainment of the ductility required to resist earthquake motion must be emphasized.

The Universal Building Code (UBC) has an entire section on earthquake construction requirements for structures by occupancy. These are the latest state of the art concepts based on nationally recognized good practices. Furthermore, building permits will not be issued until earthquake designs have been considered and building officials have approved the design or lack of in writing. For those parts of the country where snow is a major consideration, weight factors also need to be considered when preparing the station design.

The design should consider several factors to avoid adverse impacts and potential harm, and to ensure safe, stable and compatible development appropriate to site conditions. These should include, but not be limited to:

- review of available literature regarding the site and surrounding areas,
- detailed topographic analysis,
- subsurface data and exploration logs,
- ground surface profiles,
- analysis of relationship of vegetated cover and slope stability,
- site stability analysis,
- geotechnical considerations to reduce risk, and
- construction and post construction monitoring.

Properly designed and constructed modern fire and EMS stations are reasonably safe from natural disasters. In the past 25 years, seismologists, geotechnical and structural engineers have developed building design standards that allow properly rehabilitated structures to suffer relatively minor damage in an earthquake.

Prior to the 1970's, the structural design of buildings for earthquake focused primarily on providing sufficient strength to overcome the forces resulting from ground motion, The most glaring deficiency was the need for increased building ductility. Prior to that time, earthquake building codes required that structures be designed for a prescribed level of static force. An earthquake is a dynamic event, not a static one. Accordingly, structures need not only strength, but also the ability to sustain their strength for the duration of the earthquake.

1. Provide a design with continuity in the floor plan from a structural perspective.

2. Employ structures with continuous earthquake load carrying systems. Stations with wall systems that are discontinuous or interrupted by changes in the floor plan perform poorly.

   Modern buildings designed to building codes developed since the mid-1970's will generally perform well. Exceptions do exist-most notably the need for greater ductility in certain components of concrete buildings, and the need for improved design of connections in structural steel buildings. In general, however modern fire/EMS stations that have been designed and constructed properly perform well, The most serious damage potential exists in older stations or those structures in which serious design or construction errors exist.

3. Determine the level of earthquake resistance to be designed into the new station or of the degree to which an older one should be upgraded.

4. Use an importance factor of 1.5 in the station design. The importance factor is a ranking of the need for building survivability as applied to structural integrity. It relates to the specification of materials and design techniques that allow building survivability.

The Federal Emergency Management Agency has established the National Earthquake Hazard Reduction Program (NEHRP) and uses the Building Seismic Safety Council (BSSC) of the National Institute of Building Sciences for developing guidelines for evaluating/refurbishing existing facilities and for designing/building earthquake resistant structures. BSSC has developed several publications in this area:

- FEMA *172, Techniques for Seismically Rehabilitating Existing Buildings*
- FEMA *273, Guidelines for Seismic Rehabilitating Buildings*
- FEMA *274, NEHRP Commentary on Guidelines for Seismic Rehabilitating Buildings*
- FEMA *178, NEHRP Handbook for Seismic Evaluation of Existing Buildings*
- FEMA 222A, *1994 Edition of NEHRP Recommended Provisions for Seismic Regulators for New Buildings*
- FEMA *222B, Commentary, Part II, on 1994 Edition of NEHRP Recommended Provisions for Seismic Regulators for New Buildings*

The commentary publications FEMA 274 and 222B contain extensive information beyond the guidelines and recommended provisions of the respective companion publications.

Generally, the first consideration for avoiding the potential for flooding is the station's location with specific attention on the type of area soils, water table, and drainage. The Federal Insurance Administration of the Federal Emergency Management Agency together with state and local authorities establish the potential for area flooding by classifying flood risk in every region within the United States.

Flooding

If past history has demonstrated susceptibility of existing stations to flooding or significantly hindering operations, consideration should be given to relocating the station to a more desirable location, Justification for moving a station can be made on the basis of maintaining operations during natural disasters when fire and emergency medical services are often most needed in addition to the costs to the community in terms of relatively slow response times.

Design of structures for avoiding potential flood damage utilizes some of the same factors that apply to earthquake-resistant design. No only must stations be designed to prevent the inflow of water from flooding, but they must also take in account to forces of running water and pressure created by flood waters, Particularly important are:

- nearby sewer and runoff capacity,
- local area flood controls,
- eliminating basements and crawl spaces from station design,
- design of adequate footings and foundations for anchoring structures,
- allowing for appropriate subfloor drainage,
- waterproofing concrete and masonary walls,
- foundation elevation above the expected flood level, and
- use of sump pumps and other devices to reduce ground water accumulation.

FEMA Publication 102, *Flood-proofing Non-Resident Structures,* offers detailed information for building design and construction for reducing damage associated with flooding.

Station design attributes important in avoiding damage during severe storms such as those associated with hurricanes include specially reinforced walls, reinforced roofs, and structural connections. These design features generally add significantly to the cost of construction. Additional FEMA publications are being prepared to cover this area of building design.

For information on FEMA publications related to earthquake and flood resistant designs, contact the Federal Emergency Management Agency at 1-800-480-2520 and use the publication number to order.

Fire (all types)

Nature of the Hazard. An emergency services station is one occupancy that should have no excuse to burn and be constructed and maintained so as to be practically immune to fire. However, fire stations are often just as susceptible to fires as other structure. Fires can be caused at stations by ignition of flammable gases or liquids or electrical problems. Particularly dangerous are small undetected fires resulting from improper or damaged wiring which can travel behinds walls and in attic spaces. Dangerous fire conditions can also arise from the accumulation of fuel and flammable vapors in areas close to ignition sources.

Extent of Problem at the Station. Because of the nature of fire and EMS stations, several potential fire hazards can exist. Perhaps the most common hazard is leaving the kitchen equipment on during a response. Each year, nearly a dozen or more serious fires are reported at fire and EMS stations through the United States. Some of these fires have resulted in substantial damage to the stations affected and sometimes personnel injuries. A review of the cause for the majority of these fires appear to related to:

1.   Faulty or improper wiring/electrical connections,

2.   Equipment left operating during a response, and

2.   Handling of flammable substances within the station.

Relevant Regulations and Standards. The following are some the regulations and standards that apply to this area:

- locally adopted building and fire codes
- OSHA 1910.37 Means of Egress, General
- OSHA 1910.38 Employee Emergency Plans and Fire Prevention Plans
- OSHA 1910.157 Portable Fire Extinguishers
- OSHA 1910.158 Standpipe and Hose Systems
- OSHA 1910.159 Automatic Sprinkler Systems
- OSHA 1910.160 Fixed Extinguishing Systems, General
- OSHA 1910.161 Fixed Extinguishing Systems, Dry Chemical
- OSHA 1910.162 Fixed Extinguishing Systems, Gaseous Agent
- OSHA 1910.163 Fixed Extinguishing Systems, Water Spray and Foam
- OSHA 1910.164 Fire Detection Systems
- NFPA 1, *Fire Prevention Code*
- NFPA 10, *Portable Extinguishers*
- NFPA *13, Sprinkler Systems*
- NFPA *14, Standpipe Hose Systems*
- NFPA *70, National Electric Code*
- NFPA 70E, *Electrical Safety Requirements for Employee Workplaces*
- NFPA 101, *Life Safety Code*

Pertinent OSHA Regulations are provided in Appendix A under "Fire Protection" (p. A-

4) and "Means of Egress" (pp. A-5 to A-9) sections. Appendix B provides a list of federal offices in OSHA states while Appendix C provides the offices where state regulations can be obtained if in a non-OSHA state.

Specific organizations, listed in Appendix D, which provide guidance in this area include:

> AIA - American Institute of Architects
> AISC - American Institute of Steel Construction
> AISG - American Insurance Services Group
> AITC - American Institute of Timber Construction
> ASCE - American Society of Civil Engineers
> CED - Civil Engineering Data
> CSI - Construction Specifications Institute
> FGCC - Federal Geodetic Control Committee
> IRI - Industrial Risk Insurers
> NFPA - National Fire Protection Association
> VFIS - The Volunteer Firemen's Insurance Services, Inc.

Preventative Design Requirements. Many of the safety concerns related to fire risk have been covered in previous sections, specifically those addressing Electrocution/Shock Hazards, Explosions, and Hazardous Materials. Several specific concerns relative to fire protection exist at the fire or EMS station as in any large industrial or residential structure. These include systems for fire-resistive construction, detection of fires, means of extinguishment, and providing adequate egress.

Fire Resistive Construction

1. Separate apparatus and storage areas away from quarters or sleeping areas with at least one fire resistive separation.

2. Install wiring, connections, and appliance in accordance with NFPA 70, *National Electric Code* and NFPA 70E, *Electrical Safety Requirements for Employee Workplaces.* For the latter standard, Chapters 2 (Wiring Design and Protection), Chapter 3 (Wiring Methods, Components, and Equipment for General Use) and Chapter 5 (Hazardous Locations) are particularly relevant.

3. Ensure that fire/EMS station design and construction should be in accordance with NFPA 101, *Life Safety Code.* This standard addresses use of compartmentilization construction, smoke barriers, special hazard protection, and interior finishes.

Fire Detection

1. Require that all stations have both smoke and carbon monoxide detectors installed, preferably directly wired with battery back-up.

2. Ensure that fire detectors and fire detection systems are tested and adjusted as

often as needed to maintain proper reliability and operating condition.

3. Select the number, spacing and location of fire detectors as based upon design data obtained from field experience, or tests, engineering surveys, the manufacturer's recommendations, or a recognized testing laboratory listing as required by local code.

4. Provide alarm systems which are recognizable from other alarms systems at the station.

Fire Extinguishing Systems

1. Provide portable fire extinguishers and mount, locate and identify them so that they are readily accessible.

2. Use only approved portable fire extinguishers.

3. Ensure that portable fire extinguishers are maintained in a fully charged and operable condition and kept in their designated places at all times except during use.

4. Install automatic sprinkler designs which provide the necessary discharge patterns, densities, and water flow characteristics for complete coverage of designated station areas. Use a Hazard Design of Ordinary Group I in accordance with NFPA 13, *Sprinkler Systems*. The U. S. Fire Administration strongly recommends the use of sprinkler systems in the design of fire and EMS stations.

5. Sprinkler systems should be installed with consideration to maintenance, adequate water supplies, protection of piping, protection of sprinklers, sprinkler spacing, and appropriate sprinkler alarms.

6. There are three main types of fixed extinguishing systems: (1) dry chemical, (2) gaseous agent, and (3) water spray and foam. These specialized systems may be considered for working spaces where there is a high potential for fire hazards.

Means of Egress

1. Provided a sufficient number of exits consistent with expected maximum occupancy and in accordance with the local building code and NFPA 101, *Life Safety Code*.

2. Mark every exit and conspicuously indicate the routes for escape from any point.

3. When more than one exit is required from a story or room, arrange at least two of the exits remote from each other as to minimize any possibility that both may be blocked by any one fire or other emergency condition.

4.   Locate exits and exit access such that exits are readily accessible at all times.

5.   Use doors from a room to an exit or to a way of exit access of the side-hinged, swinging type.

6.   The minimum width of any way of exit access must not be less than 32 inches clear width and provide adequate headroom of at least 7 feet 6 inches.

7.   Supply each exit with an illuminated exist sign illuminated by a reliable light source giving a value of not less than 5 foot-candles on the illuminated surface.

8.   Arrange exiting from sleeping and day room areas so that at least one exit rout is available without going through the apparatus room.

9.   Prepare emergency action plans. Post floor plans which clearly show the emergency escape routes included in the emergency action plan. Color coding will aid employees in determining their route assignments.

10   Designate an emergency safe area or rendezvous point in the event of an emergency. Practice emergency evacuation.

In each of these areas, the fire and emergency medical services station should serve as the model for the community.

Theft and Burglary

Extent of Problem at the Station. As with any public facility, fire or EMS stations are subject to theft and burglary, particularly since stations are vacated during response periods without any attendant. This problem can be more severe for volunteer stations where no rotating occupants stay at the station. Problems can also arise during public functions at or outside uses of the station. The specific location of the station within the community can also pose particular risks to station occupants and contents.

Relevant Regulations and Standards. Regulations and standards in this area are typically found in local codes. Specific organizations, listed in Appendix D, which provide guidance in this area include:

- AISG - American Insurance Services Group
- IRI - Industrial Risk Insurers
- UL - Underwriters' Laboratories
- VFIS - The Volunteer Firemen's Insurance Services, Inc.

Preventative Design Requirements. Common sense in securing valuable items will help prevent theft as well as burglary. Even though the value of some items stolen from a station may be minimal, it is important that the necessary effort be made to reduce thefts of personal property and station equipment/supplies.

<u>Department Property</u>

1. If practical, place equipment in its storage area at the close of each shift. Silhouette boards for items such as tools are a good way of keeping tract of what is missing.

2. Inventory equipment frequently and at irregular intervals.

<u>Personal Property</u>

1. If at all possible, utilize single user lockers. The best locks are keyed, such as pin-tumbler padlocks with solid brass or steel bodies and hardened shackles at least 3/8 inches in diameter. The shackle should be locked at heel and toe. A high quality combination padlock, meeting the same specifications, may also be used.

2. Establish natural surveillance points in areas where personnel leave belongings.

3. Recommend that personal items be engraved with an identifiable mark.

    The majority of departments now mark their more valuable equipment in some manner. This is not done, however, for personal property. It is recommended that each fire or EMS station make an effort to mark personal property.

    Personal property should be marked with the individual's own state drivers license number. An electric engraving tool can be used to mark the metal or plastic equipment.

Vandalism and Violence

Nature of the Hazard. Parking is a consideration that needs to be closely examined. Many times, stations have been built where firefighters have had to park blocks from the station because no consideration was given to automobiles. Also the security of the parking lot needs to be considered. In some regions, the parking lot may need to be fenced to prevent vandalism to firefighters' vehicles while in other parts of the municipality, this may not a problem. Methods for dealing with civil unrest require installation of security measures at the station. An example is to install bullet proof shields over the windows.

Relevant Regulations and Standards. Guidelines in this area are typically found in local codes. Specific organizations, listed in Appendix D, which provide guidance in this area include:

- AISG - American Insurance Services Group
- IRI - Industrial Risk Insurers
- UL - Underwriters' Laboratories
- VFIS - The Volunteer Firemen's Insurance Services, Inc.

The Volunteer Firemen's Insurance Services, Inc. (VFIS) has prepared "Introduction to Basic Loss Control for the Emergency Service." Book 4, Buildings & Grounds Inspection/Security deals specifically with station issues related to theft and vandalism. VFIS can be contacted at 183 Leader Heights Road, P. O. Box 2726, York, PA (Phone: 800-233-1957).

Preventative Design Requirements. The prevention of vandalism and violence, as well as reducing theft and burglary (see previous section) is accomplished by the following measures:

1. *Use of barriers such as fences, gates, and barricades.* Enclosing areas with fences and gates prohibits entry by unauthorized personnel and vehicles which may be intent on theft, vandalism or other mischief. These characteristics may use actual barricade materials or be affected by the location terrain and landscape.

2. *Providing Adequate Lighting.* Lighting should be provided for the station perimeter, parking lots, walkways, entrances, and any sensitive areas which would permit easy vandalism or theft during hours of darkness. In addition, internal lighting should be provided to assist police personnel during times when the station is unoccupied.

3. *Installing Intrusion Sensors and Alarms.* A number of systems are available for stations based on different levels of sophistication, capabilities, and principles of operation. An alarm system is not a "security system" - merely part of an entire package which includes hardware, construction and design. Considerations for installation, and use of alarm systems include:

   - In new construction, incorporate system wiring needs into the original building plans. This ensures maximum ease of installation, concealability, and protection from tampering, and is also less expensive. Identify the desired protected areas prior to drafting of plans.

   - In existing construction, conceal all wiring within walls, behind moldings or beyond reach.

   - Conceal or make inaccessible all sensors and contacts.

   - Locate control boxes (including phone dialers) in a space such as a locked room or closet away from high traffic areas.

   - Place phone line connections and terminals inside the building. Phone line entry into the building must be inaccessible - underground, or out of reach and view, This is essential, since most alarm systems transmit the alarm signal via these phone lines.

   - Since comprehensive alarm coverage is not usually feasible, determine priority by value and desirability of contents.

- Establish the watch office, including records and safe, audio-visual, shops and medical storage rooms as priority areas.

Table 13 provides recommended specification for an alarm system.

4.   *Use of Lock and Key Systems.* Specific personnel are identified and permitted to have access to sensitive areas by the selective distribution and control of keys. Sensitive areas may include the apparatus fuel distribution locations, electrical cabinets, offices, personnel files, and some equipment areas. Issuance of keys should be minimized to promote their accountability.

5   *Choosing Construction Features* for Security. An examination of the design or survey of the existing station should identify possible unauthorized entry point through doors, windows, skylights, fire escapes, and ventilation ducts. Possible security-based features for consideration include:

- use of glass blocks or tempered glass for windows,
- tamper-proof door hinges,
- solid core exterior doors with deadbolts for anti-intrusion defense,
- bay doors with minimum sized windows,
- automatic bay door closers to secure the premises after response to an emergency,
- special locks and keys (including combination locks) security programs,
- window bars or locking shutters on windows for areas especially prone to vandalism and violence, and
- mirrors or cameras at entrance points or remote facility locations.

6.   *Adding Internal Security Features.* Doors to special rooms and cabinets inside these rooms should have locks to create defensive barriers to unauthorized personnel who have penetrated the station as a deterrent to damage. The use of safes for storing personal money and valuables may also be used. Signs such as "Department Personnel Only" and "Electrical Hazards - Do Not Enter" should be posted indicating that certain areas are off limits to unauthorized personnel.

7.   *Securing the Grounds.* Items outside the station which may be used to gain entry should be removed and properly stored. Examples include pallets or boxes for training and any foreign objects which may be used for breaking, prying, or creating fires.

The choice of security features should not jeopardize the life safety of station personnel or the general public authorized to be at the station.

Table 13. Detailed Specifications for Security System

A. *CONTROL PANEL:*

1. The power source should be a self-contained battery, or battery backup with an A.C. system. Eight hours minimum backup is needed for urban areas, and 24 hours for rural installations.
2. Control box keys should be held only by the Chief or watch commanders.
3. Local, audible alarms require resetting once they are activated. The audible alarm controls should allow the alarm to continue until manually turned off and reset, which should also guarantee investigation of the alarm cause. Automatic bell shut-down and reset timers are also available.

B. *ANNUNCIATION:*

An alarm signal must initiate a response. This is the function of annunciation. One of the following choices should be applicable to most situations. They are listed in descending order of security:

1. Proprietary terminal - the fire district is totally responsible for alarm monitoring. They provide facilities for monitoring panels and hire the personnel. All alarms are fed directly to this proprietary station. Consider the advantages of a cooperative system with a neighboring district.
2. Direct connection - alarms go directly to the local police station.
3. Commercial central station - alarms are directed to a professional monitoring service. The best of these meet U. L. standards.
4. Answering service - alarms sent by phone dialers are received by this 24 hour service.
5. Local alarms - a bell, horn or a siren or light on the premises is activated by the system with the intent of alerting neighbors and passersby, and frightening the intruder away before he has completed his objective.

C. *RESPONSE:*

1. The alarm should elicit a response by law enforcement or other designated personnel.
2. In no case should an answering service or central station provide a response, since this would require that they have a key. Maximum control is retained if fire station personnel always respond to alarms. Fire district personnel also have complete knowledge of building layout.

Table 13. Detailed Specifications for Security System (continued)

D. *BUILDING SURFACE INTRUSION DETECTORS:*

1. Only the keyed outside doors in each building need to be protected by detection contacts if non-key operated doors have been properly secured. These contacts will activate the alarm when the door is opened.
2. Station windows cannot be feasibly protected by an alarm system because of cost and vulnerability to tampering and false alarms.

E. *INTERIOR SPACE INTRUSION DETECTORS:*

Many types of detectors are available and do a good job when the right sensors are used for specific locations. It is likely that the specific needs of the various locations will require using a combination of intrusion detectors. The choice of these detectors should be made by a security specialist.

1. Commonly used types include magnetically or mechanically activated contact switches on doors or interior windows leading to valuable target areas. The criminally-wise youngster or the professional burglar can, however, bypass or break through such a protected door without activating the alarm.
2. Photoelectric beams direct an invisible beam across the protected area which, when broken, activates the alarm. Photoelectric beam protection is useful to protect entrances, exits, corridors and multiple office areas.
3. Sound monitoring systems are expensive but allow the monitoring personnel to evaluate the alarm situation because they can hear activity in the building.
4. Ultrasonic detectors detect motion within a predetermined area. Thermostatically controlled forced air heat, bells and falling banners and posters can cause false alarms.
5. Microwave detectors also detect motion but their high frequency beam will penetrate most walls and glass, thus increasing the change of outside movement causing an alarm.
6. Infrared systems also detect motion. They sense the infrared energy which all people emit. Nighttime headlights, certain heaters and window exposures to sunlight can cause false alarms.

SECTION 5 - COMPLIANCE AND FUNDING ISSUES

Determining Compliance

As this manual has shown, there are numerous safety and health issues which apply to the design or operation of a fire or EMS station. While national and state regulations exist for determining compliance with respect to many of these issues, the majority of regulations will be from local sources. Different regions of the country follow different codes. A check with the local authority (city, county or state government) will help to determine which set of codes apply in the area. In addition, there are several organizations which specialize in the development of model codes that have been adopted locally. These organizations listed below with further information in Appendix D can provide assistance in identifying appropriate requirements for the specific building area:

- BOCA - Building Officials & Code Administrators (east, northeast, and midwest);
- ICBO - International Council of Building Officials (west, southwest, northwest);
- NCSBCS - National Conference of States on Building Codes and Standards;
- SBCCI - Southern Building Code Congress International (south and southeast).

One of the most important steps in the design process following the needs assessment is to acquire the appropriate information which will let the department be aware of all pertinent regulations and alternatives for meeting the design goals. This manual has been intended to assist in this process. As specific safety and health concerns are identified in the design process, the pertinent sections of the manual should be consulted to identify relevant approaches and sources of additional information.

Another approach for ensuring compliance is to use a Compliance Check-Off List, Appendix F provides a representative (generic) check off list by area of the station. This check-off list is intended to be used in the design process, to assist in remodelling, and can be used for a safety and health inspection of existing stations. The form is set up as a simple "check-the-box" format by station area, providing space for noting specific discrepancies in designs or existing stations. As with any system, it cannot account for all specific safety and health requirements specific to each area.

Funding

A major consideration in the design or remodelling process is funding. One strong consideration is the insurance industry. These companies run smoke detector programs, scholarship programs, arson hot line reward programs, and a multitude of other programs related to the fire service. If a fire or EMS department can show a reduction in fire loss due to aggressive fire prevention, there may be a possibility that the insurance savings could go to enhancing existing facility or building new ones.

A number of alternative sources of funding are available to the fire and emergency medical services. Among these are:

- fees for special services,
- benefit assessments,
- impact development fees, and
- subscription charges.

Major local government funding mechanisms include taxes, borrowing, leasing, benefit assessment changes, fees, contracts, cost sharing, subscriptions, and impact development fees. The latter is usually involved for purchase of new fire and EMS stations and their full complement of equipment. In addition to these sources, local governments can also obtain funding from state and federal programs. Often federal program funds are distributed through state programs. Among the various sources are fire insurance surcharges, vehicle-related fees, special state grant programs, general state revenues, state-provided services, and federal grant programs. Some volunteer fire and EMS departments raise much of their funds from the private sector through direct solicitation, fund-raising events, corporate donations, private foundations, and community service clubs.

A complete description of funding alternatives is offered in "A Guide to Funding Alternatives for Fire and Emergency Medical Service Departments, U.S. Fire Administration Report FA 141, Federal Emergency Management Agency, U.S. Fire Administration, December 1993. Request a copy of this publication by writing to the Fire Management & Technical Programs Division, U. S. Fire Administration at 16825 South Seton Avenue, Emmitsburg, MD 21727.

Achieving a Safe and Healthy Work Environment

Making department stations safe and healthy requires much more than simply meeting applicable regulations. It requires a commitment from the department at all levels to ensure that the station provides an example to the community in terms of safety and health standards. It requires the attitude that stations are more than a temporary residence for emergency responders, but an asset for which communities will increasingly become dependent on in the future. **As** with any safety and health related endeavor, fire/EMS departments should focus on the following with regard to station design:

1. Explicitly determine station needs based on department/community requirements;

2. Carefully supervise the design process to ensure that needs are addressed;

3. Thoroughly review the safety and health of the station as it is being constructed and periodically after it has been built; and

4. Use "lessons learned" and experience to establish the basis for future needs.

Given sufficient funding and community support, this process will ensure that stations provide a safe and health environment for emergency responders.

APPENDIX A

OSHA REGULATIONS PERTAINING
TO FIRE/EMS STATION SAFETY AND HEALTH

APPENDIX A. OSHA REGULATIONS PERTAINING TO FIRE/EMS STATION SAFETY AND HEALTH

Section 6(a) of the Williams-Steiger Occupational Safety and Health Act of 1970 (84 Stat. 1593) provides that "without regard to chapter 5 of Title 5, United States Code, or to the other subsections of this section, the Secretary shall, as soon as practicable during the period beginning with the effective date of this Act and ending 2 years after such date, by rule promulgate as an occupational safety or health standard any national consensus standard, and any established Federal standard, unless he determines that the promulgation of such a standard would not result in improved safety or health for specifically designated employees." The legislative purpose of this provision is to establish, as rapidly as possible and without regard to the rule-making provisions of the Administrative Procedure Act, standards with which industries are generally familiar, and on whose adoption interested and affected persons have already had an opportunity to express their views. Such standards are either (1) national consensus standards on whose adoption affected persons have reached substantial agreement, or (2) Federal standards already established by Federal statutes or regulations.

> The following requirements are OSHA regulations that could influence the construction/remodel and maintenance of fire and emergency medical services stations from Part 1910 - General Occupational Safety and Health Standards. Specific regulations are organized by area of the station and by types of hazards. **This list may not contain all applicable OSHA regulations that apply to station design for safety and health. The following regulations are current through July 1, 1996.**

ORGANIZATION OF EXCERPTED REGULATIONS

FIRE PROTECTION

1910.164(c)(2) The employer shall assure that fire detectors and fire detection systems are tested and adjusted as often as needed to maintain proper reliability and operating condition except that factory calibrated detectors need not be adjusted after installation.

1910.164(c)(5) The employer shall also assure that fire detectors that need to be cleaned of dirt, dust, or other particulates in order to be fully operational are cleaned at regular periodic intervals.

1910.164(d)(3) The employer shall assure that detectors are supported independently of their attachment to wires or tubing.

1910.164(f) *Number, location and spacing of detecting devices.* The employer shall assure that the number, spacing and location of fire detectors is based upon design data obtained from field experience, or tests, engineering surveys, the manufacturer's recommendations, or a recognized testing laboratory listing.

1910.36(d)(2) Every automatic sprinkler system, fire detection and alarm system, exit lighting, fire door, and other item of equipment, where provided, shall be continuously in proper operating condition.

Fire Extinguishers

1910.157(c)(l) The employer shall provide portable fire extinguishers and shall mount, locate and identify them so that they are readily accessible to employees without subjecting the employees to possible injury.

1910.157(c)(2) Only approved portable fire extinguishers shall be used to meet the requirements of this section.

1910.157(c)(3) The employer shall not provide or make available in the workplace portable fire extinguishers using carbon tetrachloride or chlorobromomethane extinguishing agents.

1910.157(c)(4) The employer shall assure that portable fire extinguishers are maintained in a fully charged and operable condition and kept in their designated places at all times except during use.

Automatic Sprinkler Systems

1910.37(m) *Automatic sprinkler systems.* All automatic sprinkler systems shall be continuously maintained in reliable operating condition at all times, and such periodic inspections and tests shall be made as are necessary to assure proper maintenance.

1910.37(n) *Fire alarm signaling systems.* The employer shall assure that fire alarm signaling systems are maintained and tested.

1910.159(c)(l)(i) All automatic sprinkler designs used to comply with this standard shall provide the necessary discharge patterns, densities, and water flow characteristics for complete coverage in a particular workplace or zoned subdivision of the workplace.

1910.159(c)(l)(ii) The employer shall assure that only approved equipment and devices are used in the design and installation of automatic sprinkler systems used to comply with this standard.

MEANS OF EGRESS

> **NOTE:** The OSHA requirements are based on older egress requirements. Any communities which have adopted the newer life safety codes may be subject to more stringent requirements. NFPA 101 is one example of such a standard.

1910.36(b)(l) Every building or structure, new or old, designed for human occupancy shall be provided with exits sufficient to permit the prompt escape of occupants in case of fire or other emergency. The design of exits and other safeguards shall be such that reliance for safety to life in case of fire or other emergency will not depend solely on any single safeguard; additional safeguards shall be provided for life safety in case any single safeguard is ineffective due to some human or mechanical failure.

1910.36(b)(3) Every building or structure shall be provided with exits of kinds, numbers, location, and capacity appropriate to the individual building or structure, with due regard to the character of the occupancy, the number of persons exposed, the fire protection available, and the height and type of construction of the building or structure, to afford all occupants convenient facilities for escape.

1910.36(b)(5) Every exit shall be clearly visible or the route to reach it shall be conspicuously indicated in such a manner that every occupant of every building or structure who is physically and mentally capable will readily know the direction of escape from any point, and each path of escape, in its entirety, shall be so arranged or marked that the way to a place of safety outside is unmistakable. Any doorway or passageway not constituting an exit or way to reach an exit, but of such a character as to be subject to being mistaken for an exit, shall be so arranged or marked as to minimize its possible confusion with an exit and the resultant danger of persons endeavoring to escape from fire finding themselves trapped in a dead-end space, such as a cellar or storeroom, from which there is no other way out.

1910.36(b)(6) In every building or structure equipped for artificial illumination, adequate and reliable illumination shall be provided for all exit facilities.

1910.36(b)(7) In every building or structure of such size, arrangement, or occupancy that a fire may not itself provide adequate warning to occupants, fire alarm facilities shall be provided where necessary to warn occupants of the existence of fire so that they may escape, or to facilitate the orderly conduct of fire exit drills.

1910.36(b)@) Every building or structure, section, or area thereof of such size, occupancy, and arrangement that the reasonable safety of numbers of occupants may be endangered by the blocking of any single means of egress due to fire or smoke, shall have at least two means of egress remote from each other, so arranged as to minimize any possibility that both may be blocked by any one fire or other emergency conditions.

1910.37(c)(l) The capacity in number of persons per unit of exit width for approved components of means of egress shall be as follows:
(i) Level Egress Components (including Class A Ramps) 100 persons.
(ii) Inclined Egress Components (including Class B Ramps) 60 persons.
(iii) A ramp shall be designated as Class A or Class B in accordance with Table E-l

TABLE E-l

Parameter	Class A	Class B
Width	44 inches and greater	30 to 44 inches.
Slope	1 to 13/16 inches in 12 inches	13/16 to 2 inches in 12 inches.
Maximum height between landings	No limit	12 feet.

1910.37(c)(2) Means of egress shall be measured in units of exit width of 22 inches. Fractions of a unit shall not be counted, except that 12 inches added to one or more full units shall be counted as one-half a unit of exit width.

1910.37(c)(3) Units of exit width shall be measured in the clear at the narrowest point of the means of egress except that a handrail may project inside the measured width on each side not more than 5 inches and a stringer may project inside the measured width not more than 1 1/2 inches. An exit or exit access door swinging into an aisle or passageway shall not restrict the effective width thereof at any point during its swing to less than the minimum widths hereafter specified.

1910.37(d) Egress capacity and occupant load.

1910.37(d)(l) The capacity of means of egress for any floor, balcony, tier, or other occupied space shall be sufficient for the occupant load thereof. The occupant load shall be the maximum number of persons that may be in the space at any time.

1910.37(d)(2) Where exits serve more than one floor, only the occupant load of each floor considered individually need be used in computing the capacity of the exits at that floor, provided that exit capacity shall not be decreased in the direction of exit travel.

1910.37(e) *Arrangement of exits.* When more than one exit is required from a story, at least two of the exits shall be remote from each other and so arranged as to minimize any possibility that both may be blocked by any one fire or other emergency condition.

1910.37(f)(l) Exits shall be so located and exit access shall be so arranged that exits are readily accessible at all times. Where exits are not immediately accessible from an open floor area, safe and continuous passageways, aisles, or corridors leading directly to every exit and so arranged as to provide convenient access for each occupant to at least two exits by separate ways of travel, except as a single exit or limited dead ends are permitted by other provisions of this subpart, shall be maintained.

1910.37(f)(2) A door from a room to an exit or to a way of exit access shall be of the side-hinged, swinging type. It shall swing with exit travel when the room is occupied by more than 50 persons or used for a high hazard occupancy.

1910.37(f)(3) In no case shall access to an exit be through a bathroom, or other room subject to locking, except where the exit is required to serve only the room subject to locking.

1910.37(f)(4) Ways of exit access and the doors to exits to which they lead shall be so designed and arranged as to be clearly recognizable as such. Hangings or draperies shall not be placed over exit doors or otherwise so located as to conceal or obscure any exit. Mirrors shall not be placed on exit doors. Mirrors shall not be placed in or adjacent to any exit in such a manner as to confuse the direction of exit.

1910.37(f)(5) Exit access shall be so arranged that it will not be necessary to travel toward any area of high hazard occupancy in order to reach the nearest exit, unless the path of travel is effectively shielded from the high hazard location by suitable partitions or other physical barriers.

1910.37(f)(6) The minimum width of any way of exit access shall in no case be less than 28 inches. Where a single way of exit access leads to an exit, its capacity in terms of width shall be at least equal to the required capacity of the exit to which it leads. Where more than one way of exit access leads to an exit, each shall have a width adequate for the number of persons it must accommodate.

1910.37(g)(l) Access to an exit may be by means of any exterior balcony, porch, gallery, or roof that conforms to the requirements of this section.

1910.37(g)(2) Exterior ways of exit access shall have smooth, solid floors, substantially level, and shall have guards on the unenclosed sides.

1910.37(g)(3) Where accumulation of snow or ice is likely because of the climate, the exterior way of exit access shall be protected by a roof, unless it serves as the sole normal means of access to the rooms or spaces served, in which case it may be assumed that snow and ice will be regularly removed in the course of normal occupancy.

1910.37(g)(4) A permanent, reasonably straight path of travel shall be maintained over the required exterior way of exit access. There shall be no obstruction by railings, barriers, or gates that divide the open space into sections appurtenant to individual rooms, apartments, or other uses. Where the Assistant Secretary of Labor or his duly authorized representative finds the required path of travel to be obstructed by furniture or other movable objects, he may require that they be fastened out of the way or he may require that railings or other permanent barriers be installed to protect the path of travel against encroachment.

1910.37(g)(S) An exterior way of exit access shall be so arranged that there are no dead ends in excess of 20 feet. Any unenclosed exit served by an exterior way of exit access shall be so located that no part of the exit extends past a vertical plane 20 feet and one-half the required width of the exit from the end of and at right angles to the way of exit access.

1910.37(g)(6) Any gallery, balcony, bridge, porch, or other exterior exit access that projects beyond the outside wall of the building shall comply with the requirements of this section as to width and arrangement.

1910.37(h)(l) All exits shall discharge directly to the street, or to a yard, court, or other open space that gives safe access to a public way. The streets to which the exits discharge shall be of width adequate to accommodate all persons leaving the building. Yards, courts, or other open spaces to which exits discharge shall also be of adequate width and size to provide all persons leaving the building with ready access to the street.

1910.37(h)(2) Stairs and other exits shall be so arranged as to make clear the direction of egress to the street. Exit stairs that continue beyond the floor of discharge shall be interrupted at the floor of discharge by partitions, doors, or other effective means.

1910.37(i) *Headroom.* Means of egress shall be so designed and maintained as to provide adequate headroom, but in no case shall the ceiling height be less than 7 feet 6 inches nor any projection from the ceiling be less than 6 feet 8 inches from the floor.

1910.37(i) *Changes in elevation.* Where a means of egress is not substantially level, such differences in elevation shall be negotiated by stairs or ramps.

1910.37(k)(l) Doors, stairs, ramps, passages, signs, and all other components of means of egress shall be of substantial, reliable construction and shall be built or installed in a workmanlike manner.

1910.37(k)(2) Means of egress shall be continuously maintained free of all obstructions or impediments to full instant use in the case of fire or other emergency.

1910.37(k)(3) Any device or alarm installed to restrict the improper use of an exit shall be so designed and installed that it cannot, even in cases of failure, impede or prevent emergency use of such exit.

1910.37(1)(l) No furnishings, decorations, or other objects shall be so placed as to obstruct exits, access thereto, egress therefrom, or visibility thereof.

1910.37(l)(2) No furnishings or decorations of an explosive or highly flammable character shall be used in any occupancy.

1910.37(q)(l) Exits shall be marked by a readily visible sign. Access to exits shall be marked by readily visible signs in all cases where the exit or way to reach it is not immediately visible to the occupants.

1910.37(q)(2) Any door, passage, or stairway which is neither an exit nor a way of exit access, and which is so located or arranged as to be likely to be mistaken for an exit, shall be identified by a sign reading "Not an Exit" or similar designation, or shall be identified by a sign indicating its actual character, such as "To Basement," "Storeroom," "Linen Closet," or the like.

1910.37(q)(3) Every required sign designating an exit or way of exit access shall be so located and of such size, color, and design as to be readily visible. No decorations, furnishings, or equipment which impair visibility of an exit sign shall be permitted, nor shall there be any brightly illuminated sign (for other than exit purposes), display, or object in or near the line of vision to the required exit. sign of such a character as to so detract attention from the exit sign that it may not be noticed.

1910.37(q)(4) Every exit sign shall be distinctive in color and shall provide contrast with decorations, interior finish, or other signs.

1910.37(q)(5) A sign reading "Exit", or similar designation, with an arrow indicating the directions, shall be placed in every location where the direction of travel to reach the nearest exit is not immediately apparent.

1910.37(q)(6) Every exit sign shall be suitably illuminated by a reliable light source giving a value of not less than 5 foot-candles on the illuminated surface. Artificial lights giving illumination to exit signs other than the internally illuminated types shall have screens, discs, or lenses of not less than 25 square inches area made of translucent material to show red or other specified designating color on the side of the approach.

1910.37(q)(7) Each internally illuminated exit sign shall be provided in all occupancies where reduction of normal illumination is permitted.

1910.37(q)(8) Every exit sign shall have the word "Exit" in plainly legible letters not less than 6 inches high, with the principal strokes of letters not less than three-fourths-inch wide.

ALL AREAS - ELECTRICAL

General

1910.303(b)(1) *Examination.* Electrical equipment shall be free from recognized hazards that are likely to cause death or serious physical harm to employees. Safety of equipment shall be determined using the following considerations:

1910.303(b)(1)(i) Suitability for installation and use in conformity with the provisions of this subpart, Suitability of equipment for an identified purpose may be evidenced by listing or labeling for that identified purpose.

1910.303(b)(1)(ii) Mechanical strength and durability, including, for parts designed to enclose and protect other equipment, the adequacy of the protection thus provided.

1910.303(b)(2) *Installation and* use. Listed or labeled equipment shall be used or installed in accordance with any instructions included in the labeling.

1910.303(e) *Marking.* Electrical equipment may not be used unless the manufacturer's name, trademark, or other descriptive marking by which the organization responsible for the product may be identified is placed on the equipment. Other markings shall be provided giving voltage, current, wattage, or other ratings as necessary. The marking shall be of sufficient durability to withstand the environment involved.

1910.303(f) Identification of disconnecting means and circuits. Each disconnecting means required by this subpart for motors and appliances shall be legibly marked to indicate its purpose, unless located and arranged so the purpose is evident. Each service, feeder, and branch circuit, at its disconnecting means or overcurrent device, shall be legibly marked to indicate its purpose, unless located and arranged so the purpose is evident. These markings shall be of sufficient durability to withstand the environment involved.

1910.303(g)(1) *Working space about electric equipment.* Sufficient access and working space shall be provided and maintained about all electric equipment to permit ready and safe operation and maintenance of such equipment.

1910.303(g)(1)(iii) *Access and entrance to working space.* At least one entrance of sufficient area shall be provided to give access to the working space about electric equipment.

1910.303(g)(1)(vi) **Headroom.** The minimum headroom of working spaces about service equipment, switchboards, panel-boards, or motor control centers shall be 6 feet 3 inches.

1910.303(h)(4)(ii) Permanent ladders or stairways shall be provided to give safe access to the working space around electric equipment installed on platforms, balconies, mezzanine floors, or in attic or roof rooms or spaces.

Wiring Design and Protection

1910.304(a)(1) *Identification of conductors.* A conductor used as a grounded conductor shall be identifiable and distinguishable from all other conductors.

1910.304(a)(2) *Polarity of connections.* No grounded conductor may be attached to any terminal or lead so as to reverse designated polarity.

1910.304(a)(3) *Use of grounding terminals and devices.* A grounding terminal or grounding-type device on a receptacle, cord connector, or attachment plug may not be used for purposes other than grounding.

1910.304(c)(3) *Clearance from building openings.* Conductors shall have a clearance of at least 3 feet from windows, doors, porches, fire escapes, or similar locations. Conductors run above the top level of a window are considered to be out of reach from that window and, therefore, do not have to be 3 feet away.

1910.304(c)(4) *Clearance over roofs.* Conductors shall have a clearance of not less than 8 feet from the highest point of roofs over which they pass, except that:

1910.304(d)(1)(i) *General.* Means shall be provided to disconnect all conductors in a building or other structure from the service-entrance conductors. The disconnecting means shall plainly indicate whether it is in the open or closed position and shall be installed at a readily accessible location nearest the point of entrance of the service-entrance conductors.

1910.304(e)(1)(vi)(A) Circuit breakers shall clearly indicate whether they are in the open (off) or closed (on) position.

1910,304(e)(1)(vi)(B) Where circuit breaker handles on switchboards are operated vertically rather than horizontally or rotationally, the up position of the handle shall be the closed (on) position.

1910.304(f)(1)(i) All 3-wire DC systems shall have their neutral conductor grounded.

1910.304(f)(4) *Grounding path.* The path to ground from circuits, equipment, and enclosures shall be permanent and continuous.

1910.304(f)(5)(i) *Supports and enclosures for conductors.* Metal cable trays, metal raceways, and metal enclosures for conductors shall be grounded, except that:

1910.304(f)(5)(i)(A) Metal enclosures such as sleeves that are used to protect cable assemblies from physical damage need not be grounded;

1910.304(f)(5)(i)(B) Metal enclosures for conductors added to existing installations of open wire, knob-and-tube wiring, and nonmetallic-sheathed cable need not be grounded if all of the following conditions are met: (1) Runs are less than 25 feet; (2) enclosures are free from probable contact with ground, grounded metal, metal laths, or other conductive materials; and enclosures are guarded against employee contact.

1910,304(f)(5)(iii) *Frames of ranges and clothes dryers.* Frames of electric ranges, wall-mounted ovens, counter-mounted cooking units, clothes dryers, and metal outlet or junction boxes which are part of the circuit for these appliances shall be grounded.

1910.304(f)(5)(v) *Equipment connected by cord and plug.* Under any of the conditions described in paragraphs (f)(5)(v)(A) through (f)(5)(v)(C) of this section, exposed non-current-carrying metal parts of cord - and plug-connected equipment which may become energized shall be grounded. If the equipment is of the following types:

(C)(1) Refrigerators, freezers, and air conditioners;

(C)(2) Clothes-washing, clothes-drying and dishwashing machines, sump pumps, and electrical aquarium equipment;

(C)(3) Hand-held motor-operated tools;

(C)(4) Motor-operated appliances of the following types: hedge clippers, lawn mowers, snow blowers, and wet scrubbers;

(C)(5) Cord- and plug-connected appliances used in damp or wet locations or by employees standing on the ground or on metal floors or working inside of metal tanks or boilers;

(C)(6) Portable and mobile X-ray and associated equipment;

(C)(7) Tools likely to be used in wet and conductive locations; and

(C)(8) Portable hand lamps.

Wiring Methods, Components, and Equipment General Use

1910,305(c)(2) *Faceplates for flush-mounted snap switches.* Flush snap switches that are mounted in ungrounded metal boxes and located within reach of conducting floors or other conducting surfaces shall be provided with faceplates of nonconducting, noncombustible material.

1910.305(d) *Switchboards and panelboards.* Switchboards that have any exposed live parts shall be located in permanently dry locations and accessible only to qualified persons. Panelboards shall be mounted in cabinets, cutout boxes, or enclosures approved for the purpose and shall be dead front. However, panelboards other than the dead front externally-operable type are permitted where accessible only to qualified persons. Exposed blades of knife switches shall be dead when open.

1910.305(e)(1) Cabinets, cutout boxes, fittings, boxes, and panelboard enclosures in damp or wet locations shall be installed so as to prevent moisture or water from entering and accumulating within the enclosures. In wet locations the enclosures shall be weatherproof.

1910.305(e)(2) Switches, circuit breakers, and switchboards installed in wet locations shall be enclosed in weatherproof enclosures.

1910.305(f) *Conductors for general wiring.* All conductors used for general wiring shall be insulated unless otherwise permitted. The conductor insulation shall be of a type that is approved for the voltage, operating temperature, and location of use. Insulated conductors shall be distinguishable by appropriate color or other suitable means as being grounded conductors, ungrounded conductors, or equipment grounding conductors.

1910,305(g)(l)(i) Flexible cords and cables shall be approved and suitable for conditions of use and location. Flexible cords and cables shall be used only for:
(A) Pendants;
(B) Wiring of fixtures;
(C) Connection of portable lamps or appliances;
(D) Elevator cables;
(E) Wiring of cranes and hoists;
(F) Connection of stationary equipment to facilitate their frequent interchange;
(G) Prevention of the transmission of noise or vibration;
(H) Appliances where the fastening means and mechanical connections are designed to permit removal for maintenance and repair; or
(1) Data processing cables approved as a part of the data processing system.

1910.305(g)(l)(iii) Unless specifically permitted, flexible cords and cables may not be used:
(A) As a substitute for the fixed wiring of a structure;
(B) Where run through holes in walls, ceilings, or floors;
(C) Where run through doorways, windows, or similar openings;
(D) Where attached to building surfaces; or
(E) Where concealed behind building walls, ceilings, or floors.

1910.305(g)(2)(iii) Flexible cords shall be connected to devices and fittings so that strain relief is provided which will prevent pull from being directly transmitted to joints or terminal screws,

1910.305(j)(2)(i) Receptacles, cord connectors, and attachment plugs shall be constructed so that no receptacle or cord connector will accept an attachment plug with a different voltage or current rating than that for which the device is intended. However, a 20-ampere T-slot receptacle or cord connector may accept a 15-ampere attachment plug of the same voltage rating.

1910.305(j)(2)(ii) A receptacle installed in a wet or damp location shall be suitable for the location.

1910.305(j)(5)(ii) The operating voltage of exposed live parts of transformer installations shall be indicated by warning signs or visible markings on the equipment or structure.

1910.305(j)(5)(vi) Transformer vaults shall be constructed so as to contain fire and combustible liquids within the vault and to prevent unauthorized access. Locks and latches shall be so arranged that a vault door can be readily opened from the inside.

Specific Purpose Equipment and Installations

1910.306(c)(l) *Disconnecting means.* Elevators, dumbwaiters, escalators, and moving walks shall have a single means for disconnecting all ungrounded main power supply conductors for each unit.

1910.306(h)(9)(i) The conductive surfaces of cranes and hoists that enter the cell line working zone need not be grounded. The portion of an overhead crane or hoist which contacts an energized electrolytic cell or energized attachments shall be insulated from ground.

1910.306(j)(1) This section apply to electric wiring for and equipment in or adjacent to all swimming, wading, therapeutic, and decorative pools and fountains, whether permanently installed or storable, and to metallic auxiliary equipment, such as pumps, filters, and similar equipment.

1910.306(j)(2)(i) *Receptacles.* A single receptacle of the locking and grounding type that provides power for a permanently installed swimming pool recirculating pump motor may be located not less than 5 feet from the inside walls of a pool. All other receptacles on the property shall be located at least 10 feet from the inside walls of a pool. Receptacles which are located within 15 feet of the inside walls of the pool shall be protected by ground-fault circuit interrupters.

Note: In determining these dimensions, the distance to be measured is the shortest path the supply cord of an appliance connected to the receptacle would follow without piercing a floor, wall, or ceiling of a building or other effective permanent barrier.

1910.306(j)(2)(ii)(A) Unless they are 12 feet above the maximum water level, lighting fixtures and lighting outlets may not be installed over a pool or over the area extending 5 feet horizontally from the inside walls of a pool. However, a lighting fixture or lighting outlet which has been installed before April 16, 1981, may be located less than 5 feet measured horizontally from the inside walls of a pool if it is at least 5 feet above the surface of the maximum water level and shall be rigidly attached to the existing structure. It shall also be protected by a ground-fault circuit interrupter installed in the branch circuit supplying the fixture.

1910.306(j)(2)(ii)(B) Unless installed 5 feet above the maximum water level and rigidly attached to the structure adjacent to or enclosing the pool, lighting fixtures and lighting outlets installed in the area extending between 5 feet and 10 feet horizontally from the inside walls of a pool shall be protected by a ground-fault circuit interrupter.

1910.306(j)(3)(i) Cord- and plug-connected lighting fixtures installed within 16 feet of the water surface of permanently installed pools.

1910.306(j)(4)(i) A ground-fault circuit interrupter shall be installed in the branch circuit supplying underwater fixtures operating at more than 15 volts. Equipment installed underwater shall be approved for the purpose.

1910.306(j)(4)(ii) No underwater lighting fixtures may be installed for operation at over 150 volts between conductors.

1910.306(j)(5) *Fountains.* All electric equipment operating at more than 15 volts, including power supply cords, used with fountains shall be protected by ground-fault circuit interrupters,

LIVING SPACES - GENERAL

Sanitation

1910.141(b)(1)(i) Potable water shall be provided in all places of employment, for drinking, washing of the person, cooking, washing of foods, washing of cooking or eating utensils, washing of food preparation or processing premises, and personal service rooms.

1910.141(b)(1)(iii) Portable drinking water dispensers shall be designed, constructed, and serviced so that sanitary conditions are maintained, shall be capable of being closed, and shall be equipped with a tap.

1910.141(b)(2)(i) Outlets for non-potable water, such as water for industrial or firefighting purposes, shall be posted or otherwise marked in a manner that will indicate clearly that the water is unsafe and is not to be used for drinking, washing of the person, cooking, washing of food, washing of cooking or eating utensils, washing of food preparation or processing premises, or personal service rooms, or for washing clothes.

1910.141(b)(2)(ii) Construction of nonpotable water systems or systems carrying any other nonpotable substance shall be such as to prevent backflow or backsiphonage into a potable water system.

1910.141(c)(1)(i) Toilet rooms separate for each sex, shall be provided in all places of employment in accordance with Table J-l of this section. The number of facilities to be provided for each sex shall be based on the number of employees of that sex for whom the facilities are furnished. Where toilet rooms will be occupied by no more than one person at a time, can be locked from the inside, and contain at least one water closet, separate toilet rooms for each sex need not be provided. Where such single-occupancy rooms have more than one toilet facility, only one such facility in each toilet room shall be counted for the purpose of Table J-l.

TABLE J-1

Number of Employees	Minimum Number of Water Closets (1)
1-15	1
16-35	2
36-55	3
56-80	4
81-110	5
111-150	6
Over 150	(2)

Footnotes: (1) Where toilet facilities will not be used by women, urinals may be provided instead of water closets, except that the number of water closets in such cases shall not be reduced to less than 2/3 of the minimum specified.

(2) 1 additional fixture for each additional 40 employees.

1910.141(c)(1)(iii) The sewage disposal method shall not endanger the health of employees.

1910.141(c)(2)(i) Each water closet shall occupy a separate compartment with a door and walls or partitions between fixtures sufficiently high to assure privacy.

1910.141(d)(1) Washing facilities shall be maintained in a sanitary condition.

1910.141(d)(2)(ii) Each lavatory shall be provided with hot and cold running water, or tepid running water.

1910.141(d)(2)(iii) Hand soap or similar cleansing agents shall be provided.

1910.141(d)(2)(iv) Individual hand towels or sections thereof, of cloth or paper, warm air blowers or clean individual sections of continuous cloth toweling, convenient to the lavatories, shall be provided.

1910.141(d)(3)(ii) One shower shall be provided for each 10 employees of each sex, or numerical fraction thereof, who are required to shower during the same shift.

1910.141(d)(3)(iii) Body soap or other appropriate cleansing agents convenient to the showers shall be provided.

1910.141(d)(3)(iv) Showers shall be provided with hot and cold water feeding a common discharge line.

1910.141(d)(3)(v) Employees who use showers shall be provided with individual clean towels.

1910.141(e) *Change rooms.* Whenever employees are required by a particular standard to wear protective clothing because of the possibility of contamination with toxic materials, change rooms equipped with storage facilities for street clothes and separate storage facilities for the protective clothing shall be provided.

1910.141(f) *Clothes drying facilities.* Where working clothes are provided by the employer and become wet or are washed between shifts, provision shall be made to insure that such clothing is dry before reuse.

1910.141(g)(2) *Eating and drinking areas.* No employee shall be allowed to consume food or beverages in a toilet room nor in any area exposed to a toxic material.

1910.141(g)(3) *Waste disposal containers.* Receptacles constructed of smooth, corrosion resistant, easily cleanable, or disposable materials, shall be provided and used for the disposal of waste food. The number, size, and location of such receptacles shall encourage their use and not result in overfilling. They shall be emptied not less frequently than once each working day, unless unused, and shall be maintained in a clean and sanitary condition. Receptacles shall be provided with a solid tight-fitting cover unless sanitary conditions can be maintained without use of a cover.

1910.141(g)(4) *Sanitary storage. No* food or beverages shall be stored in toilet rooms or in an area exposed to a toxic material.

1910.141(h) *Food handling.* All employee food service facilities and operations shall be carried out in accordance with sound hygienic principles. In all places of employment where all or part of the food service is provided, the food dispensed shall be wholesome, free from spoilage, and shall be processed, prepared, handled, and stored in such a manner as to be protected against contamination.

Illumination

1910.265(c)(g)(i) *Adequacy.* Illumination shall be provided and designed to supply adequate general and local lighting to rooms, buildings, and work areas during the time of use.

1910.305(9)(iii)(f) Lamps for general illumination shall be protected from accidental contact or breakage. Protection shall be provided by elevation of at least 7 feet from normal working surface or by a suitable fixture or lampholder with a guard.

1910.303(g)(1)(v) *Illumination.* Illumination shall be provided for all working spaces about service equipment, switchboards, panelboards, and motor control centers installed indoors.

1910.305(j)(1)(i) Fixtures, lampholders, lamps, rosettes, and receptacles may have no live parts normally exposed to employee contact. However, rosettes and cleat-type lampholders and receptacles located at least 8 feet above the floor may have exposed parts.

1910.305(i)(1)(ii) Handlamps of the portable type supplied through flexible cords shall be equipped with a handle of molded composition or other material approved for the purpose, and a substantial guard shall be attached to the lampholder or the handle.

1910.305(j)(1)(iii) Lampholders of the screw-shell type shall be installed for use as lampholders only. Lampholders installed in wet or damp locations shall be of the weatherproof type.

1910.305(j)(1)(iv) Fixtures installed in wet or damp locations shall be approved for the purpose and shall be so constructed or installed that water cannot enter or accumulate in wireways, lampholders, or other electrical parts.

1910.261(b)(7) *Emergency illumination.* Where emergency lighting is necessary, the system shall be so arranged that the failure of any individual lighting element, such as the burning out of a light.

Noise

1910.95(a) Protection against the effects of noise exposure shall be provided when the sound levels exceed those shown in Table G-16 when measured on the A scale of a standard sound level meter at slow response.

1910.95(b)(1) When employees are subjected to sound exceeding those listed in Table A-X. feasible administrative or engineering controls shall be utilized. If such controls fail to reduce sound levels within the levels of Table G-16. personal protective equipment shall be provided and used to reduce sound levels within the levels of the table.

1910.95(b)(2) If the variations in noise level involve maxima at intervals of 1 second or less, it is to be considered continuous.

TABLE G-16. PERMISSIBLE NOISE EXPOSURES

Duration per day, hours	Sound level dBA slow response
8	90
6	92
4	95
3	97
2	100
1½	102
1	105
½	110
¼ or less	115

SPECIFIC INTERIOR AREAS

Kitchen

1910.141(b)(1)(i) Potable water shall be provided in all places of employment, for drinking, washing of the person, cooking, washing of foods, washing of cooking or eating utensils, washing of food preparation or processing premises, and personal service rooms.

1910.304(f)(5)(iii) *Frames of ranges and clothes dryers.* Frames of electric ranges, wall-mounted ovens, counter-mounted cooking units, clothes dryers, and metal outlet or junction boxes which are part of the circuit for these appliances shall be grounded.

1910.304(f)(5)(v)(C)(1) Refrigerators, freezers, and air conditioners; shall be grounded.

1910,304(f)(5)(v)(C)(2) Clothes-washing, clothes-drying and dishwashing machines, sump pumps, and electrical aquarium equipment; shall be grounded

Laundry

1910.262(c)(1) *Interlocking device.* Each drying tumbler, each double cylinder shaker or clothes tumbler, and each washing machine shall be equipped with an interlock device which will prevent the power operation of the inside cylinder when the outer door on the case or shell is open, and which will also prevent the outer door on the case or shell from being opened without shutting off the power.

1910.264(c)(1)(ii)(b) Each washing machine shall be provided with means for holding open the doors or covers of inner and outer cylinders or shells while being loaded or unloaded.

Medical-Related Areas

1910.151(b) In the absence of an infirmary, clinic, or hospital in near proximity to the workplace which is used for the treatment of all injured employees, a person or persons shall be adequately trained to render first aid. First aid supplies approved by the consulting physician shall be readily available.

1910.1030(c)(1)(i) *Exposure Control Plan.* Each employer having an employee(s) with occupational exposure as defined by paragraph (b) of this section shall establish a written Exposure Control Plan designed to eliminate of minimize employee exposure.

1910.1030(c)(1)(ii) The Exposure Control Plan shall contain at least the following elements:
(A) The exposure determination required by paragraph (c)(2),
(B) The schedule and method of implementation for paragraphs (d) Methods of Compliance, (e) HIV and HBV Laboratories and Production Facilities, (f) Hepatitis B Vaccination and Post-Exposure Evaluation and Follow-up, (g) Communication of Hazards to Employees, and (h) Recordkeeping, of this standard, and
(C) The procedure for the evaluation of circumstances surrounding exposure incidents as required by paragraph (f)(3)(i) of this standard.

1910.1030(c)(1)(iii) Each employer shall ensure that a copy of the Exposure Control Plan is accessible to employees in accordance with 29 CFR 1910.20(e).

1910.1030(c)(1)(iv) The Exposure Control Plan shall be reviewed and updated at least annually and whenever necessary to reflect new or modified tasks and procedures which affect occupational exposures and to reflect new or revised employee positions with occupational exposure.

1910,1030(c)(1)(v) The Exposure Control Plan shall be made available to the Assistant Secretary and the Director upon request for examination and copying.

1910.1030(d)(2)(i) Engineering and work practice controls shall be used to eliminate or minimize employee exposure. Where occupational exposure remains after institution of these controls, personal protective equipment shall also be used.

1910.1030(d)(2)(iii) Employers shall provide handwashing facilities which are readily accessible to employees.

1910.1030(d)(2)(viii) Immediately or as soon as possible after use, contaminated reusable sharps shall be placed in appropriate containers until properly reprocessed. These containers shall be:
(A) puncture resistant;
(B) labeled or color-coded in accordance with this standard;
(C) leakproof on the sides and bottom

1910.1030(d)(4)(ii)(E) Reusable sharps that are contaminated with blood or other potentially infectious materials shall not be stored or processed in a manner that requires employees to reach by hand into the containers where these sharps have been placed.

1910.1030(d)(3)(iii) *Accessibility.* The employer shall ensure that appropriate personal protective equipment in the appropriate sizes is readily accessible at the worksite or is issued to employees. Hypoallergenic gloves, glove liners, powderless gloves, or other similar alternatives shall be readily accessible to those employees who are allergic to the gloves normally provided.

1910.1030(d)(3)(iv) *Cleaning, Laundering, and Disposal.* The employer shall clean, launder, and dispose of personal protective equipment at no cost to the employee.

1910.1030(d)(3)(viii) When personal protective equipment is removed it shall be placed in an appropriately designated area or container for storage, washing, decontamination or disposal.

1910.1030(d)(4)(i) *General.* Employers shall ensure that the worksite is maintained in a clean and sanitary condition. The employer shall determine and implement an appropriate written schedule for cleaning and method of decontamination based upon the location within the facility, type of surface to be cleaned, type of soil present, and tasks or procedures being performed in the area.

1910.1030(d)(4)(II) All equipment and environmental and working surfaces shall be cleaned and decontaminated after contact with blood or other potentially infectious materials.

1910.1030(d)(4)(ii)(C) All bins, pails, cans, and similar receptacles intended for reuse which have a reasonable likelihood for becoming contaminated with blood or other potentially infectious materials shall be inspected and decontaminated on a regularly scheduled basis and cleaned and decontaminated immediately or as soon as feasible upon visible contamination.

1910.1030(d)(4)(iii)(C) Disposal of all regulated waste shall be in accordance with applicable regulations of the United States, States and Territories, and political subdivisions of States and Territories.

1910.1030(d)(4)(iv)(A)(1) Contaminated laundry shall be bagged or containerized at the location where it was used and shall not be sorted or rinsed in the location of use.

1910.1030(e)(4)(i) The work areas shall be separated from areas that are open to unrestricted traffic flow within the building.Passage through two sets of doors shall be the basic requirement for entry into the work area from access corridors or other contiguous areas. Physical separation of the high-containment work area from access corridors or other areas or activities may also be provided by a double-doored clothes-change room (showers may be included), airlock, or other access facility that requires passing through two sets of doors before entering the work area.

1910.1030(e)(4)(ii) The surfaces of doors, walls, floors and ceilings in the work area shall be water resistant so that they can be easily cleaned. Penetrations in these surfaces shall be sealed or capable of being sealed to facilitate decontamination.

1910.1030(e)(4)(iii) Each work area shall contain a sink for washing hands and a readily available eye wash facility.The sink shall be foot, elbow, or automatically operated and shall be located near the exit door of the work area.

1910.1030(e)(4)(iv) Access doors to the work area or containment module shall be self-closing.

1910.1030(e)(4)(vi) A ducted exhaust-air ventilation system shall be provided. This system shall create directional airflow that draws air into the work area through the entry area, exhaust air shall not be recirculated to any other area of the building, shall be discharged to the outside, and shall be dispersed away from occupied areas and air intakes. The proper direction of the airflow shall be verified (i.e., into the work area).

INTERIOR AREAS - WALKING-WORKING SURFACES

General

1910.22(a)(1) All places of employment, passageways, storerooms, and service rooms shall be kept clean and orderly and in a sanitary condition.

1910.22(a)(2) The floor of every workroom shall be maintained in a clean and, so far as possible, a dry condition. Where wet processes are used, drainage shall be maintained, and false floors, platforms, mats, or other dry standing places should be provided where practicable.

1910.22(b)(1) Where mechanical handling equipment is used, sufficient safe clearances shall be allowed for aisles, at loading docks, through doorways and wherever turns or passage must be made. Aisles and passageways shall be kept clear and in good repairs, with no obstruction across or in aisles that could create a hazard.

1910.22(b)(2) Permanent aisles and passageways shall be appropriately marked.

1910.22(c) *Covers and guardrails.* Covers and/or guardrails shall be provided to protect personnel from the hazards of open pits, tanks, vats, ditches, etc.

1910.22(d)(1) In every building or other structure, or part thereof, used for mercantile, business, industrial, or storage purposes, the loads approved by the building official shall be marked on plates of approved design which shall be supplied and securely affixed by the owner of the building, or his duly authorized agent, in a conspicuous place in each space to which they relate. Such plates shall not be removed or defaced but, if lost, removed, or defaced, shall be replaced by the owner or his agent.

1910.23(a)(1) Every stairway floor opening shall be guarded by a standard railing constructed in accordance with paragraph (e) of this section. The railing shall be provided on all exposed sides (except at entrance to stairway). For infrequently used stairways where traffic across the opening prevents the use of fixed standard railing (as when located in aisle spaces, etc.), the guard shall consist of a hinged floor opening cover of standard strength and construction and removable standard railings on all exposed sides (except at entrance to stairway).

1910.36(d)(2) Every automatic sprinkler system, fire detection and alarm system, exit lighting, fire door, and other item of equipment, where provided, shall be continuously in proper operating condition.

1910.23(a)(3) Every hatchway and chute floor opening shall be guarded by one of the following:

1910.23(a)(3)(i) Hinged floor opening cover of standard strength and construction equipped with standard railings or permanently attached thereto so as to leave only one exposed side. When the opening is not in use, the cover shall be closed or the exposed side shall be guarded at both top and intermediate positions by removable standard railings.

1910.23(a)(3)(ii) A removable railing with toeboard on not more than two sides of the opening and fixed standard railings with toeboards on all other exposed sides. The removable railings shall be kept in place when the opening is not in use.

1910.23(a)(4) Every skylight floor opening and hole shall be guarded by a standard skylight screen or a fixed standard railing on all exposed sides.

1910.23(a)(5) Every pit and trapdoor floor opening, infrequently used, shall be guarded by a floor opening cover of standard strength and construction. While the cover is not in place, the pit or trap opening shall be constantly attended by someone or shall be protected on all exposed sides by removable standard railings.

1910.23(a)(6) Every manhole floor opening shall be guarded by a standard manhole cover which need not be hinged in place. While the cover is not in place, the manhole opening shall be constantly attended by someone or shall be protected by removable standard railings.

1910.23(a)(8) Every floor hole into which persons can accidentally walk shall be guarded by either:
(i) A standard railing with standard toeboard on all exposed sides, or

(ii) A floor hole cover of standard strength and construction. While the cover is not in place, the floor hole shall be constantly attended by someone or shall be protected by a removable standard railing.

1910.23(a)(9) Every floor hole into which persons cannot accidentally walk (on account of fixed machinery, equipment, or walls) shall be protected by a cover that leaves no openings more than 1 inch wide. The cover shall be securely held in place to prevent tools or materials from falling through.

1910.23(a)(10) Where doors or gates open directly on a stairway, a platform shall be provided, and the swing of the door shall not reduce the effective width to less than 20 inches.

1910.23(b)(1) Every wall opening from which there is a drop of more than 4 feet shall be guarded by one of the following:

1910.23(b)(1)(i) Rail, roller, picket fence, half door, or equivalent barrier. Where there is exposure below to falling materials, a removable toe board or the equivalent shall also be provided. When the opening is not in use for handling materials, the guard shall be kept in position regardless of a door on the opening. In addition, a grab handle shall be provided on each side of the opening with its center approximately 4 feet above floor level and of standard strength and mounting.

1910.23(b)(1)(ii) Extension platform onto which materials can be hoisted for handling, and which shall have side rails or equivalent guards of standard specifications.

1910.23(b)(2) Every chute wall opening from which there is a drop of more than 4 feet shall be guarded by one or more of the barriers specified in paragraph (b)(l) of this section or as required by the conditions.

1910.23(b)(3) Every window wall opening at a stairway landing, floor, platform, or balcony, from which there is a drop of more than 4 feet, and where the bottom of the opening is less than 3 feet above the platform or landing, shall be guarded by standard slats, standard grill work (as specified in paragraph (e) (11) of this section), or standard railing.

1910.23(b)(5) Where there is a hazard of materials falling through a wall hole, and the lower edge of the near side of the hole is less than 4 inches above the floor, and the far side of the hole more than 5 feet above the next lower level, the hole shall be protected by a standard toeboard, or an enclosing screen either of solid construction, or as specified in paragraph (e) (11) of this section.

1910.23(c)(l) Every open-sided floor or platform 4 feet or more above adjacent floor or ground level shall be guarded by a standard railing (or the equivalent as specified in paragraph (e) (3) of this section) on all open sides except where there is entrance to a ramp, stairway, or fixed ladder. The railing shall be provided with a toeboard wherever, beneath the open sides,

1910.23(c)(2) Every runway shall be guarded by a standard railing (or the equivalent as specified in paragraph (e)(3) of this section) on all open sides 4 feet or more above floor or ground level. Wherever tools, machine parts, or materials are likely to be used on the runway, a toeboard shall also be provided on each exposed side.

1910.23(c)(3) Regardless of height, open-sided floors, walkways, platforms, or runways above or adjacent to dangerous equipment, pickling or galvanizing tanks, degreasing units, and similar hazards shall be guarded with a standard railing and toe board.

1910.23(d)(1) Every flight of stairs having four or more risers shall be equipped with standard stair railings or standard handrails as specified in as not to constitute a projection hazard.

1910.23(e)(5)(ii) The height of handrails shall be not more than 34 inches nor less than 30 inches from upper surface of handrail to surface of tread in line with face of riser or to surface of ramp.

1910.23(e)(5)(iii) The size of handrails shall be: When of hardwood, at least 2 inches in diameter; when of metal pipe, at least 1 1/2 inches in diameter. The length of brackets shall be such as will give a clearance between handrail and wall or any projection thereon of at least 3 inches. The spacing of brackets shall not exceed 8 feet.

1910.23(e)(5)(iv) The mounting of handrails shall be such that the completed structure is capable of withstanding a load of at least 200 pounds applied in any direction at any point on the rail.

1910.23(e)(6) All handrails and railings shall be provided with a clearance of not less than 3 inches between the handrail or railing and any other object.

1910.23(e)(7) Floor opening covers may be of any material that meets the following strength requirments:

1910.23(e)(7)(i) Trench or conduit covers and their supports, when located in plant roadways, shall be designed to carry a truck rear-axle load of at least 20,000 pounds.

1910.23(e)(7)(ii) Manhole covers and their supports, when located in plant roadways, shall comply with local standard highway requirements if any; otherwise, they shall be designed to carry a truck rear-axle load of at least 20,000 pounds.

1910.23(e)(7)(iii) The construction of floor opening covers may be of any material that meets the strength requirements. Covers projecting not more than 1 inch above the floor level may be used providing all edges are chamfered to an angle with the horizontal of not over 30 degrees. All hinges, handles, bolts, or other parts shall set flush with the floor or cover surface.

1910.23(e)(8) Skylight screens shall be of such construction and mounting that they are capable of withstanding a load of at least 200 pounds applied perpendicularly at any one area on the screen. They shall also be of such construction and mounting that under ordinary loads or impacts, they will not deflect downward sufficiently to break the glass below them. The construction shall be of grillwork with openings not more than 4 inches long or of slatwork with openeings not more than 2 inches wide with length unrestricted.

1910.23(e)(9) Wall opening barriers (rails, rollers, picket fences, and half doors) shall be of such construction and mounting that, when in place at the opening, the barrier is capable of withstanding a load of at least 200 pounds applied in any direction (except upward) at any point on the top rail or corresponding member.

1910.23(e)(10) Wall opening grab handles shall be not less than 12 inches in length and shall be so mounted as to give 3 inches clearance from the side framing of the wall opening. The size, material, and anchoring of the grab handle shall be such that the completed structure is capable of withstanding a load of at least 200 pounds applied in any direction at any point of the handle.

1910.23(e)(11) Wall opening screens shall be of such construction and mounting that they are capable of withstanding a load of at least 200 pounds applied horizontally at any point on the near side of the screen. They may be of solid construction, of grillwork with openings not more than 8 inches long, or of slatwork with openings not more than 4 inches wide with length unrestricted.

1910.24(f) *Stair treads.* All treads shall be reasonably slip-resistant and the nosings shall be of nonslip finish. Welded bar grating treads without nosings are acceptable providing the leading edge can be readily identified by personnel descending the stairway and provided the tread is serrated or is of definite nonslip design. Rise height and tread width shall be uniform throughout any flight of stairs including any foundation structure used as one or more treads of the stairs.

1910.23(d)(1) Every flight of stairs having four or more risers shall be equipped with standard stair railings or standard handrails.

1910.23(d)(1)(i) On stairways less than 44 inches wide having both sides enclosed, at least one handrail, preferably on the right side descending.

1910.23(d)(1)(iv) On stairways more than 44 inches wide but less than 88 inches wide, one handrail on each enclosed side and one stair railing on each open side.

1910.23(d)(1)(v) On stairways 88 or more inches wide, one handrail on each enclosed side, one stair railing on each open side, and one intermediate stair railing located approximately midway of the width.

1910.23(d)(2) Winding stairs shall be equipped with a handrail offset to prevent walking on all portions of the treads having width less than 6 inches.

1910.24(i) *Vertical clearance.* Vertical clearance above any stair tread to an overhead obstruction shall be at least 7 feet measured from the leading edge of the tread.

1910.265(c)(5)(ii) *Handrails.* Stairways shall be provided with a standard handrail on at least one side or on any open side. Where stairs are more than four feet wide there shall be a standard handrail at each side, and where more than eight feet wide, a third standard handrail shall be erected in the center of the stairway.

1910.265(c)(6)(i) *Opening.* Doors shall not open directly on or block a flight of stairs, and shall swing in the direction of exit travel.

1910.265(c)(6)(ii) *Stairway railings and handrails.* Every flight of stairs having four or more risers shall be equipped with standard stair railings or standard handrails as follows:

Fixed Stairs (Hose Towers)

1910.24(b) Fixed stairs shall be provided for access from one structure level to another where operations necessitate regular travel between levels, and for access to operating platforms at any equipment which requires attention routinely during operations. Fixed stairs shall also be provided where access to elevations is daily or at each shift for such purposes as gauging, inspection, regular maintenance, etc., where such work may expose employees to acids, caustics, gases, or other harmful substances, or for which purposes the carrying of tools or equipment by hand is normally required. (It is not the intent of this section to preclude the use of fixed ladders for access to elevated tanks, towers, and similar structures, overhead traveling cranes, etc., where the use of fixed ladders is common practice.) Spiral stairways shall not be permitted except for special limited usage and secondary access situations where it is not practical to provide a conventional stairway. Winding stairways may be installed on tanks and similar round structures where the diameter of the structure is not less than five (5) feet.

1910.24(c) *Stair strength.* Fixed stairways shall be designed and constructed to carry a load of five times the normal live load anticipated but never of less strength than to carry safely a moving concentrated load of 1,000 pounds.

1910.24(d) *Stair width.* Fixed stairways shall have a minimum width of 22 inches.

1910.24(e) *Angle of stairway* rise. Fixed stairs shall be installed at angles to the horizontal of between 30 deg. and 50 deg. Any uniform combination of rise/tread dimensions may be used that will result in a stairway at an angle to the horizontal within the permissible range. Table D-1 gives rise/tread dimensions which will produce a stairway within the permissible range, stating the angle to the horizontal produced by each combination. However, the rise/tread combinations are not limited to those given in Table D-l.

1910.24(f) *Stair treads.* All treads shall be reasonably slip-resistant and the nosings shall be of nonslip finish. Welded bar grating treads without nosings are acceptable providing the leading edge can be readily identified by personnel descending the stairway and provided the tread is serrated or is of definite nonslip design. Rise height and tread width shall be uniform throughout any flight of stairs including any foundation structure used as one or more treads of the stairs.

1910.24(g) *Stairway platforms.* Stairway platforms shall be no less than the width of a stairway and a minimum of 30 inches in length measured in the direction of travel.

1910.24(h) *Railings and handrails.* Standard railings shall be provided on the open sides of all exposed stairways and stair platforms. Handrails shall be provided on at least one side of closed stairways preferably on the right side descending.

Table D-1

Angle to horizontal	Rise (in inches)	Tread run (in inches)
30 deg.x35'	6 1/2	11
32 deg.x08'	6 3/4	10 3/4
33 deg.x41'	7	10 1/2
35 deg.x16'	7 1/4	10 1/4
36 deg.x52'	7 1/2	10
38 deg.x29'	7 3/4	9 3/4
40 deg.x08'	8	9 1/2
41 deg.x44'	8 1/4	9 1/4
43 deg.x22'	8 1/2	9
45 deg.x00'	8 3/4	8 3/4
46 deg.x38'	9	8 1/2
48 deg.x16'	9 1/4	8 1/4
49 deg.x54'	9 1/2	8

1910.24(i) *Vertical clearance.* Vertical clearance above any stair tread to an overhead obstruction shall be at least 7 feet measured from the leading edge of the tread.

1910.66(f)(5)(ii)(J) Access to and egress from a working platform, except for those that land directly on a safe surface, shall be provided by stairs, ladders, platforms and runways. Access gates shall be self-closing and self-latching.

Fixed Ladders (Hose Towers)

1910.27(a)(1) *Design considerations.* All ladders, appurtenances, and fastenings shall be designed to meet the following load requirements:

1910.27(a)(1)(i) The minimum design live load shall be a single concentrated load of 200 pounds.

1910.27(a)(1)(ii) The number and position of additional concentrated live-load units of 200 pounds each as determined from anticipated usage of the ladder shall be considered in the design.

1910.27(a)(1)(iii) The live loads imposed by persons occupying the ladder shall be considered to be concentrated at such points as will cause the maximum stress in the structural member being considered.

1910.27(a)(1)(iv) The weight of the ladder and attached appurtenances together with the live load shall be considered in the design of rails and fastenings.

1910.27(b)(1)(i) All rungs shall have a minimum diameter of three-fourths inch for metal ladders.

1910.27(b)(1)(ii) The distance between rungs, cleats, and steps shall not exceed 12 inches and shall be uniform throughout the length of the ladder.

1910.27(b)(1)(iii) The minimum clear length of rungs or cleats shall be 16 inches.

1910.27(b)(1)(iv) Rungs, cleats, and steps shall be free of splinters, sharp edges, burrs, or projections which may be a hazard.

1910.27(b)(1)(v) The rungs of an individual-rung ladder shall be so designed that the foot cannot slide off the end.

1910.27(b)(2) *Side rails.* Side rails which might be used as a climbing aid shall be of such cross sections as to afford adequate gripping surface without sharp edges, splinters, or burrs.

1910.27(b)(3) *Fastenings.* Fastenings shall be an integral part of fixed ladder design.

1910.27(b)(5) *Electrolytic action.* Adequate means shall be employed to protect dissimilar metals from electrolytic action when such metals are joined.

1910.27(b)(6) *Welding.* All welding shall be in accordance with the "Code for Welding in Building Construction" (AWSDl.0-1966).

1910.27(b)(7)(i) Metal ladders and appurtenances shall be painted or otherwise treated to resist corrosion and rusting when location demands. Ladders formed by individual metal rungs imbedded in concrete, which serve as access to pits and to other areas under floors, are frequently located in an atmosphere that causes corrosion and rusting. To increase rung life in such atmosphere, individual metal rungs shall have a minimum diameter of 1 inch or shall be painted or otherwise treated to resist corrosion and rusting.

1910.27(b)(7)(iii) When different types of materials are used in the construction of a ladder, the materials used shall be so treated as to have no deleterious effect one upon the other.

1910.27(c)(1) *Climbing side.* On fixed ladders, the perpendicular distance from the centerline of the rungs to the nearest permanent object on the climbing side of the ladder shall be 36 inches for a pitch of 76 degrees, and 30 inches for a pitch of 90 degrees, with minimum clearances for intermediate pitches varying between these two limits in proportion to the slope, except as provided in subparagraphs (3) and (5) of this paragraph.

1910.27(c)(2) *Ladders without cages or wells.* A clear width of at least 15 inches shall be provided each way from the centerline of the ladder in the climbing space, except when cages or wells are necessary.

1910.27(c)(4) *Clearance in buck* of *ladder.* The distance from the centerline of rungs, cleats, or steps to the nearest permanent object in back of the ladder shall be not less than 7 inches, except that when unavoidable obstructions are encountered

1910.27(c)(5) *Clearance in buck of grab bar.* The distance from the centerline of the grab bar to the nearest permanent object in back of the grab bars shall be not less than 4 inches. Grab bars shall not protrude on the climbing side beyond the rungs of the ladder which they serve.

1910.27(c)(6) *Step-across distance.* The step-across distance from the nearest edge of ladder to the nearest edge of equipment or structure shall be not more than 12 inches, or less than 2 1/2 inches.

1910.27(c)(7) *Hatch cover.* Counterweighted hatch covers shall open a minimum of 60 degrees from the horizontal. The distance from the centerline of rungs or cleats to the edge of the hatch opening on the climbing side shall be not less than 24 inches for offset wells or 30 inches for straight wells. There shall be not protruding potential hazards within 24 inches of the centerline of rungs or cleats; any such hazards within 30 inches of the centerline of the rungs or cleats shall be fitted with deflector plates placed at an angle of 60 degrees from the horizontal as indicated in figure.

1910.27(d)(1)(ii) Cages or wells shall be provided on ladders of more than 20 feet to a maximum unbroken length of 30 feet.

1910.27(d)(1)(iii) Cages shall extend a minimum of 42 inches above the top of landing, unless other acceptable protection is provided.

1910.27(d)(1)(iv) Cages shall extend down the ladder to a point not less than 7 feet nor more than 8 feet above the base of the ladder, with bottom flared not less than 4 inches, or portion of cage opposite ladder shall be carried to the base.

1910.27(d)(1)(v) Cages shall not extend less than 27 nor more than 28 inches from the centerline of the rungs of the ladder. Cage shall not be less than 27 inches in width. The inside shall be clear of projections. Vertical bars shall be located at a maximum spacing of 40 degrees around the circumference of the cage; this will give a maximum spacing of approximately 9 1/2 inches, center to center.

1910.27(d)(1)(vi) Ladder wells shall have a clear width of at least 15 inches measured each way from the centerline of the ladder. Smooth-walled wells shall be a minimum of 27 inches from the centerline of rungs to the well wall on the climbing side of the ladder. Where other obstructions on the climbing side of the ladder exist, there shall be a minimum of 30 inches from the centerline of the rungs.

1910.27(d)(2) *Landing platforms.* When ladders are used to ascend to heights exceeding 20 feet (except on chimneys), landing platforms shall be provided for each 30 feet of height or fraction thereof, except that, where no cage, well, or ladder safety device is provided, landing platforms shall be provided for each 20 feet of height or fraction thereof. Each ladder section shall be offset from adjacent sections. Where installation conditions (even for a short, unbroken length) require that adjacent sections be offset, landing platforms shall be provided at each offset.

1910.27(d)(2)(i) Where a man has to step a distance greater than 12 inches from the centerline of the rung of a ladder to the nearest edge of structure or equipment, a landing platform shall be provided. The minimum step-across distance shall be 2 1/2 inches.

1910.27(d)(2)(ii) All landing platforms shall be equipped with standard railings and toeboards, so arranged as to give safe access to the ladder. Platforms shall be not less than 24 inches in width and 30 inches in length.

1910.27(d)(2)(iii) One rung of any section of ladder shall be located at the level of the landing laterally served by the ladder. Where access to the landing is through the ladder, the same rung spacing as used on the ladder shall be used from the landing platform to the first rung below the landing.

1910.27(d)(3) *Ladder extensions.* The side rails of through or side-step ladder extensions shall extend 3 1/2 feet above parapets and landings. For through ladder extensions, the rungs shall be omitted from the extension and shall have not less than 18 nor more than 24 inches clearance between rails. For side-step or offset fixed ladder sections, at landings, the side rails and rungs shall be carried to the next regular rung beyond or above the 3 1/2 feet minimum (fig. D-10).

1910.27(d)(5) *Ladder safety devices.* Ladder safety devices may be used on tower, water tank, and chimney ladders over 20 feet in unbroken length in lieu of cage protection, No landing platform is required in these cases. All ladder safety devices such as those that incorporate lifebelts, friction brakes, and sliding attachments shall meet the design requirements of the ladders which they serve.

Sliding Poles

1910.23(a)(8) Every floor hole into which persons can accidently walk shall be guarded by either:
(i) A standard railing with standard toeboard on all exposed sides; or
(ii) A floor hole cover of standard strength and construction. While the cover is not in place, the floor hole shall be constantly attended by someone or shall be protective by a removable standard railing.

1910.265(c)(9)(i) *Adequacy.* Illumination shall be provided and designed to supply adequate general and local lighting to rooms, buildings, and work areas during the time of use.

1910.261(b)(7) *Emergency illumination.* Where emergency lighting is necessary, the system shall be so arranged that the failure of any individual lighting element, such as the burning out of a light

EQUIPMENT, TOOLS AND MAINTENANCE

Batteries

1910.178(g)(2) Facilities shall be provided for flushing and neutralizing spilled electrolyte, for fire protection, for protecting charging apparatus from damage by trucks, and for adequate ventilation for dispersal of fumes from gassing batteries.

1910.178(g)(4) A conveyor, overhead hoist, or equivalent material handling equipment shall be provided for handling batteries.

1910.305(j)(7) *Storage batteries.* Provisions shall be made for sufficient diffusion and ventilation of gases from storage batteries to prevent the accumulation of explosive mixtures.

1910.151(c) Where the eyes or body of any person may be exposed to injurious corrosive materials, suitable facilities for quick drenching or flushing of the eyes and body shall be provided within the work area for immediate emergency use.

1910.102(a) *Acetylene Cylinders.* The in-plant transfer, handling, storage, and utilization of acetylene in cylinders shall be in accordance with Compressed Gas Association Pamphlet G-l-1966.

Jacks

1910.244(a)(1)(i) The operator shall make sure that the jack used has a rating sufficient to lift and sustain the load.

1910.244(a)(1)(ii) The rated load shall be legibly and permanently marked in a prominent location on the jack by casting, stamping, or other suitable means.

1910.244(a)(2)(vi) Each jack shall be thoroughly inspected at times which depend upon the service conditions. Inspections shall be not less frequent than the following:
(a) For constant and intermittent use at one locality, once every 6 months,
(b) For jacks used out of shop for special work, when sent out and when returned,
(c) For a jack subjected to abnormal load or shock, immediately before and immediately thereafter.

Servicing Multi-piece and Single Piece Wheels

1910.177(d)(3)(i) Each restraining device or barrier shall have the capacity to withstand the maximum force that would be transferred to it during a rim wheel separation occurring at 150

percent of the maximum tire specification pressure for the type of rim wheel being serviced.

1910.177(d)(5) Current charts or rim manuals containing instructions for the type of wheels being serviced shall be available in the service area.

Cranes

1910.179(b)(2) *New and existing equipment.* All new overhead and gantry cranes constructed and installed on or after August 31, 1971, shall meet the design specifications of the American National Standard Safety Code for Overhead and Gantry Cranes, ANSI B30.2.0-1967.

Power Tools

1910.212(a)(3)(iv) The following are some of the machines which usually require point of operation guarding:
(d) Power presses.
(e) Milling machines.
(f) Power saws.
(h) Portable power tools.

1910.212(a)(4) *Barrels, containers, and drums.* Revolving drums, barrels, and containers shall be guarded by an enclosure which is interlocked with the drive mechanism, so that the barrel, drum, or container cannot revolve unless the guard enclosure is in place.

1910.212(a)(5) *Exposure of blades.* When the periphery of the blades of a fan is less than seven (7) feet above the floor or working level, the blades shall be guarded. The guard shall have openings no larger than one-half (1/2) inch.

1910.212(b) *Anchoring fixed machinery.* Machines designed for a fixed location shall be securely anchored to prevent walking or moving.

1910.243(a)(2)(i) All hand-held gasoline powered chain saws shall be equipped with a constant pressure throttle control that will shut off the power to the saw chain when the pressure is released.

1910.242(b) *Compressed air used for cleaning.* Compressed air shall not be used for cleaning purposes except where reduced to less than 30 p.s. i. and then only with effective chip guarding and personal protective equipment.

Abrasive Wheel Machinery

1910.215(a)(2) *Guard design.* The safety guard shall cover the spindle end, nut, and flange projections. The safety guard shall be mounted so as to maintain proper alignment with the wheel, and the strength of the fastenings shall exceed the strength of the guard.

1910.215(a)(4) *Work rests.* On offhand grinding machines, work rests shall be used to support the work. They shall be of rigid construction and designed to be adjustable to compensate for wheel wear. Work rests shall be kept adjusted closely to the wheel with a maximum opening of one-eighth inch to prevent the work from being jammed between the wheel and the rest, which may cause wheel breakage. The work rest shall be securely clamped after each adjustment. The adjustment shall not be made with the wheel in motion.

1910.215(b)(3) *Bench and floor stands.* The angular exposure of the grinding wheel periphery and sides for safety guards used on machines known as bench and floor stands should not exceed 90 deg. or one-fourth of the periphery. This exposure shall begin at a point not more than 65 deg. above the horizontal plane of the wheel spindle.

Compressed Gases

1910.101(a) *Inspection of compressed gas cylinders.* Each employer shall determine that compressed gas cylinders under his control are in a safe condition to the extent that this can be determined by visual inspection. Visual and other inspections shall be conducted as prescribed in the Hazardous Materials Regulations of the Department of Transportation (49 CFR parts 171-179 and 14 CFR part 103). Where those regulations are not applicable, visual and other inspections shall be conducted in accordance with Compressed Gas Association Pamphlets C-6-1968 and C-8-1962.

1910.101(b) *Compressed gases.* The in-plant handling, storage, and utilization of all compressed gases in cylinders, portable tanks, rail tankcars, or motor vehicle cargo tanks shall be in accordance with Compressed Gas Association Pamphlet P-1-1965.

1910.101(c) *Safety relief devices for compressed gas containers.* Compressed gas cylinders, portable tanks, and cargo tanks shall have pressure relief devices installed and maintained in accordance with Compressed Gas Association Pamphlets S-l. l-1963 and 1965 addenda and S-1.2-1963.

Hazard Materials

1910.1200(e)(1) *Hazard Communication Program.* Employers shall develop, implement, and maintain at each workplace, a written hazard communication program which at least describes how the criteria specified in paragraphs (f), (g), and (h) of this section for labels and other forms of warning, material safety data sheets, and employee information and training will be met, and which also includes the following:

1910.1200(e)(1)(i) A list of the hazardous chemicals known to be present using an identity that is referenced on the appropriate material safety data sheet (the list may be compiled for the workplace as a whole or for individual work areas); and,

1910.1200(e)(1)(ii) The methods the employer will use to inform employees of the hazards of non-routine tasks (for example, the hazards associated with chemicals contained in unlabeled pipes in their work areas.

1910.1200(f)(l) *Labelling.* The chemical manufacturer, importer, or distributor shall ensure that each container of hazardous chemicals leaving the workplace is labeled, tagged or marked with the following information:
(i) Identity of the hazardous chemical(s);
(ii) Appropriate hazard warnings; and

(iii) Name and address of the chemical manufacturer, importer, or other responsible party.

1910.1200(f)(3) Chemical manufacturers, importers, or distributors shall ensure that each container of hazardous chemicals leaving the workplace is labeled, tagged, or marked in accordance with this section in a manner which does not conflict with the requirements of the Hazardous Materials Transportation Act (49 U. S .C. 1801 et seq.) And regulations issued under that Act by the Department of Transportation.

1910.1200(f)(4) If the hazardous chemical is regulated by OSHA in a substance-specific health standard, the chemical manufacturer, importer, distributor or employer shall ensure that the labels or other forms of warning used are in accordance with the requirements of that standard.

1910.1200(f)(7) The employer is not required to label portable containers into which hazardous chemicals are transferred from labeled containers, and which are intended only for the immediate use of the employee who performs the transfer. For purposes of this section, drugs which are dispensed by a pharmacy to a health care provider for direct administration to a patient are exempted from labeling.

1910.1200(f)(8) The employer shall not remove or deface existing labels on incoming containers of hazardous chemicals, unless the container is immediately marked with the required information.

1910.1200(f)(9) The employer shall ensure that labels or other forms of warning are legible, in English, and prominently displayed on the container, or readily available in the work area throughout each work shift. Employers having employees who speak other languages may add the information in their language to the material presented, as long as the information is presented in English as well.

1910.1200(f)(10) The chemical manufacturer, importer, distributor or employer need not affix new labels to comply with this section if existing labels already convey the required information.

1910.1200(g)(1) *Material safety data sheets.* Chemical manufactures and importers shall obtain or develop a material safety data sheet for each hazardous chemical they produce or import. Employers shall have a material safety data sheet in the workplace for each hazardous chemical which they use.

APPARATUS AREA

1910.22(a)(1) All places of employment, passageways, storerooms, and service rooms shall be kept clean and orderly and in a sanitary condition.

1910.22(a)(2) The floor of every workroom shall be maintained in a clean and, so far as possible, a dry condition. Where wet processes are used, drainage shall be maintained, and false floors, platforms, mats, or other dry standing places should be provided where practicable.

1910.22(b)(1) Where mechanical handling equipment is used, sufficient safe clearances shall be allowed for aisles, at loading docks, through doorways and wherever turns or passage must be made. Aisles and passageways shall be kept clear and in good repairs, with no obstruction across or in aisles that could create a hazard.

1910,22(b)(2) Permanent aisles and passageways shall be appropriately marked.

1910.144(a)(3) Yellow. Yellow shall be the basic color for designating caution and for marking physical hazards such as: Striking against, stumbling, falling, tripping, and "caught in between."

1910.22(d)(1) In every building or other structure, or part thereof, used for mercantile, business, industrial, or storage purposes, the loads approved by the building official shall be marked on plates of approved design which shall be supplied and securely affixed by the owner of the building, or his duly authorized agent, in a conspicuous place in each space to which they relate. Such plates shall not be removed or defaced but, if lost, removed, or defaced, shall be replaced by the owner or his agent.

1910.22(d)(2) It shall be unlawful to place, or cause, or permit to be placed, on any floor or roof of a building or other structure a load greater than that for which such floor or roof is approved by the building official.

1910.94(d)(3) *Ventilation.* Where ventilation is used to control potential exposures to workers, it shall be adequate to reduce the concentration of the air contaminant to the degree that a hazard to the worker does not exist. Methods of ventilation are discussed in American National Standard Fundamentals Governing the Design and Operation of Local Exhaust Systems, 29.2-1960.

GROUNDS/MAINTENANCE

Power Lawnmowers

1910.243(e)(l)(i) Power lawnmowers of the walk-behind, riding-rotary, and reel power lawnmowers designed for sale to the general public shall meet the design specifications in "American National Standard Safety Specifications for Power Lawnmowers" ANSI B71. l-X1968. These specifications do not apply to a walk-behind mower which has been converted to a riding mower by the addition of a sulky. Also, these specifications do not apply to flail mowers, sicklebar mowers, or mowers designed for commercial use.

1910.243(e)(2)(i) The mower blade shall be enclosed except on the bottom and the enclosure shall extend to or below the lowest cutting point of the blade in the lowest blade position.

1910.243(e)(2)(ii)(a) Warning instructions shall be affixed to the mower near the opening stating that the mower shall not be used without either the catcher assembly or the guard in place.

1910.243(e)(2)(ii)(b) The catcher assembly or the guard shall be shipped and sold as part of the mower.

1910.243(e)(2)(ii)(c) The instruction manual shall state that the mower shall not be used without either the catcher assembly or the guard in place.

1910.243(e)(2)(v) The word "Caution" or stronger wording, shall be placed on the mower at or near each discharge opening.

1910.243(e)(3)(iv) The mower handle shall be fastened to the mower so as to prevent loss of control by unintentional uncoupling while in operation.

1910.243(e)(3)(v) A positive upstop or latch shall be provided for the mower handle in the normal operating position(s). The upstop shall not be subject to unintentional disengagement during normal operation of the mower. The upstop or latch shall not allow the center or the handle grips to come closer than 17 inches horizontally behind the closest path of the mower blade(s) unless manually disengaged.

Riding Rotary Mowers

1910.243(e)(4)(i) The highest point(s) of all openings in the blade enclosure, front shall be limited by a vertical angle of opening of 15 deg. and a maximum distance of 1 1/4 inches above the lowest cutting point of the blade in the lowest blade position.

1910.243(e)(4)(ii) Opening(s) shall be placed so that grass or debris will not discharge directly toward any part of an operator seated in a normal operator position.

1910.243(e)(4)(iii) There shall be one of the following at all openings in the blade enclosure intended for the discharge of grass:

1910.243(e)(4)(iii)(a) A minimum unobstructed horizontal distance of 6 inches from the end of the discharge chute to the blade tip circle.

1910.243(e)(4)(iii)(b) A rigid bar fastened across the discharge opening, secured to prevent removal without the use of tools. The bottom of the bar shall be no higher than the bottom edge of the blade enclosure.

1910.243(e)(4)(iv) Mowers shall be provided with stops to prevent jackknifing or locking of the steering mechanism.

1910.243(e)(4)(v) Vehicle stopping means shall be provided.

1910.243(e)(4)(vi) Hand-operated wheel drive disengaging controls shall move opposite to the direction of vehicle motion in order to disengage the drive. Foot-operated wheel drive disengaging controls shall be depressed to disengage the drive. Deadman controls, both hand and foot operated, shall automatically interrupt power to a drive when the operator's actuating force is removed, and may operate in any direction to disengage the drive.

1910.304(f)(5)(v)(C)(4) Motor-operated appliances of the following types: hedge clippers, lawn mowers, snow blowers, and wet scrubbers; shall be grounded.

1910.243(a)(2)(i) All hand-held gasoline powered chain saws shall be equipped with a constant pressure throttle control that will shut off the power to the saw chain when the pressure is released.

EXTERIOR AREAS

Parking Areas

1910.144(a)(3) Yellow. Yellow shall be the basic color for designating caution and for marking physical hazards such as: Striking against, stumbling, falling, tripping, and "caught in between. "

1910.304(c)(5) *Locution of outdoor lumps.* Lamps for outdoor lighting shall be located below all live conductors, transformers, or other electric equipment, unless such equipment is controlled by a disconnecting means that can be locked in the open position or unless adequate

1910.304(f)(5)(v)(C)(4) Motor-operated appliances of the following types: hedge clippers, lawn mowers, snow blowers, and wet scrubbers; shall be grounded.

Refueling Areas

1910.106(g)(8) There shall be no smoking or open flames in the areas used for fueling, servicing fuel systems for internal combustion engines, receiving or dispensing of flammable or combustible liquids. Conspicuous and legible signs prohibiting smoking shall be posted within sight of the customer being served. The motors of all equipment being fueled shall be shut off during the fueling operation.

1910.110(h)(12) There shall be no smoking on the driveway of service stations in the dispensing areas or transport truck unloading areas. Conspicuous signs prohibiting smoking shall be posted within sight of the customer being served. Letters on such signs shall be not less than 4 inches high. The motors of all vehicles being fueled shall be shut off during the fueling operations.

1910.307(b) *Electrical installations.* Equipment, wiring methods, and installations of equipment in hazardous (classified) locations shall be intrinsically safe, approved for the hazardous (classified) location, or safe or for the hazardous (classified) location. Requirements for each of these options are as follows:

1910.307(b)(1) *Intrinsically safe.* Equipment and associated wiring approved as intrinsically safe shall be permitted in any hazardous (classified) location for which it is approved.

1910.178(f)(2) The storage and handling of liquefied petroleum gas fuel shall be in accordance with NFPA Storage and Handling of Liquefied Petroleum Gases (NFPA No. 58-1969).

Marine Areas

1910.106(a)(22) Marine service station shall mean that portion of a property where flammable or combustible liquids used as fuels are stored and dispensed from fixed equipment on shore, piers, wharves, or floating docks into the fuel tanks of self-propelled craft, and shall include all facilities used in connection therewith.

1910.106(g)(4)(i)(b) Dispensing shall be by approved dispensing units with or without integral pumps and may be located on open piers, wharves, or floating docks or on shore or on piers of the solid fill type.

1910.261(c)(3)(i) Ladders and gangplanks with railings to boat docks shall meet the requirements of American National Standards A12.1-1967, A14.1-1968, A14.2-1956, and A14.3-1956, and shall be securely fastened in place.

1910.265(c)(1) *Safety factor.* All buildings, docks, tramways, walkways, log dumps, and other structures shall be designed, constructed and maintained so as to support the imposed load in accordance with a safety factor.

1910.265(c)(4)(i) *Width.* Walkways, docks, and platforms shall be of sufficient width to provide adequate passage and working areas.

1910.265(c)(4)(iii) *Docks.* Docks and runways used for the operation of lift trucks and other vehicles shall have a substantial guard or shear timber except where loading and unloading are being performed.

APPENDIX B

REGIONAL AND AREA OSHA OFFICES

APPENDIX B. REGIONAL AND AREA OSHA OFFICES

Region or Area	Address	Phone Number
Region I (Connecticut, Maine Massachusetts, New Hampshire, Rhode Island, and Vermont)		
Boston Regional Office	133 Portland Street 1st Floor, Boston, MA 02114	617-565-7164 x150
North Boston Area Office	13 Branch Street 1st Floor Methuen, MA 01844	617-565-8110
Springfield Area Office	1145 Main Street Room 108 Springfield, MA 01103	413-785-0123
South Boston Area Office	639 Granite Street 4th Floor Braintree, MA 02184	617-565-6924
Augusta Area Office	40 Western Avenue Room 121 Augusta, ME 04330	207-622-8417
Concord Area Office	279 Pleasant Street Suite 201 Concord, NH 03301	603-225-1629
Providence Area Office	380 Westminster Mall Room 243 Providence, RI 02903	401-528-4669
Bridgeport Area Office	1 Lafayette Square Suite 202 Bridgeport, CT 06604	203-579-5579
Bangor Area Office	202 Harlow Street Room 211 Bangor, ME 00401	315-423-5188
Region II (New Jersey, New York, and Puerto Rico)		
New York Regional Office	201 Varick Street Room 670 New York, NY 10014	212-337-2378

Region or Area	Address	Phone Number
Manhattan Area Office	90 Church Street Room 1405 New York. NY 10007	212-264-9840
Syracuse Area Office	3300 Vickery Road North Syracuse, NY 13212	315-451-0808
Long Island Area Office	990 Westbury Road Westbury, NY 11590	516-334-3344
Buffalo Area Office	5360 Genesee Street Bowmansville, NY 14026	716-684-3891
Bayside Area Office	42-40 Bell Boulevard Bayside, NY 11361	718-279-906
Puerto Rico Area Office	Carlos Chardon Avenue Room 555 Hato Rey, PR 00918	809-766-5457
Albany Area Office	401 New Karner Road Suite 300 Albany, NY 12205-3809	518-464-6742
Hasbrouck Heights Area Office	500 Route 17, South 2nd Floor Hasbrouck Heights, NJ 07604	201-288-1700
Marlton Area Office	Building 2, Suite 120 701 Route 73, South Marlton, NJ 08053	609-757-5181
Avenel Area Office	Plaza 35, Suite 205 1030 St. Georges Avenue Avenel, NJ 07001	201-750-3270
Parsippany Area Office	299 Cherry Hill Road Suite 304 Parsippany, NJ 07054	201-263-1003
Tarrytown Area Office	660 White Plains Road 4th Floor Tarrytown, NY 10591	914-682-6151

APPENDIX B. REGIONAL AND AREA OSHA OFFICES (Continued)

Region or Area	Address	Phone Number
Region III (Delaware, District of Columbia, Maryland, Pennsylvania, Virginia, and West Virginia)		
Philadelphia Regional Office	Gateway Building Suite 2100 3535 Market Street Philadelphia, PA 19104	215-596-1201
Phildadelphia Area Office	U.S. Custom House Room 242 Second & Chestnut Street Philadelphia, PA 19106	215-597-4955
Wilkes - Barre Area Office	Penn Place Room 2995 20 N. Pennsylvania Avenue Wilkes-Barre, PA 18701	717-826-6538
Wilmington District Office	1 Rodney Square, Suite 402 920 King Street Wilmington, DE 19801	302-573-6115
Allentown Area Office	850 N. 5th Street Allentown, PA 18102	215-776-0592
Pittsburgh Area Office	Federal Building Room 1428 1000 Liberty Avenue Pittsburgh, PA 15222	412-644-2903
Baltimore Area Office	300 W. Pratt Street Suite 280 Baltimore, MD 21201	410-962-2840
Charleston Area Office	550 Eagan Street Room 206 Charleston, WV 25301	304-347-5937
Region IV (Alabama, Florida, Georgia, Kentucky, Mississippi, North Carolina, South Carolina, and Tennessee)		
Atlanta Regional Office	1375 Peachtree Street, N.E. Suite 587 Atlanta, GA 30367	404-347-3573

APPENDIX B. REGIONAL AND AREA OSHA OFFICES (Continued)

Region or Area	Address	Phone Number
Atlanta-East Area Office	Building 7, Suite 100 LaVista Perimeter Office Park Tucker, GA 30084	404-493-6644
Columbia Area Office	1835 Assembly Street Room 1468 Columbia, SC 29201	803-765-5904
Savannah District Office	450 Mall Boulevard Suite J Savannah, GA 31406	912-652-4383
Jackson Area Office	3780 I-55 North Suite 210 Jackson, MS	601-965-4606
Birmingham Area Office	Todd Mall 2047 Canyon Road Birmingham, AL 35216	
Mobile District Office	3737 Government Blvd Suite 100 Mobile, AL 36693	205-441-6131
Atlanta-West Area Office	LaVista Perimeter Office Park Building 7, Suite 100 Tucker, GA 30084	404-493-6644
Fort Lauderdale Area Office	Jacaranda Executive Court 8040 Peters Road Building H - 100 Ft. Lauderdale, FL 33324	305-424-0242
Jacksonville Area Office	Rebolt Building Suite 227 1851 Executive Center Dr. Jacksonville, FL 32207	904-232-2895
Frankfort Area Office	John C. Watts Federal Bldg. 330 West Broadway Room 108 Frankfort, KY 40601	502-227-7024

APPENDIX B. REGIONAL AND AREA OSHA OFFICES (Continued)

Region or Area	Address	Phone Number
Nashville Area Office	2002 Richard Jones Road Suite C-205 Nashville, TN 37215	615-781-5423
Raleigh Area Office	Century Station 300 Fayetteville Street Mall Room 438 Raleigh, NC 27601	919-856-4770
Tampa Area Office	5807 Breckenridge Parkway Suite A Tampa, FL 33610	813-626-1177
Region V (Indiana, Illinois, Michigan, Minnesota, Ohio, and Wisconsin)		
Chicago Regional Office	230 S. Dearborn Street Room 3244 Chicago, IL 60604	312-353-2220
Calumet City Area Office	1600 167th Street Suite 12 Calumet City, IL 60409	708-891-3800
Cleveland Area Office	Federal Office Building Room 899 1240 E. 9th Street Cleveland, OH 44199	216-522-3818
Chicago North Area Office	2360 E. Devon Avenue Suite 1010 Des Plaines, IL 60018	708-803-4800
Columbus Area Office	Federal Office Building Room 620 200 N. High Street Columbus, OH 43215	614-469-5582
Aurora Area Office	344 Smoke Tree Business Park NorthAurora, IL 60542	708-896-8700
Indianapolis Area Office	46 E. Ohio Street Room 423 Indianapolis, IN 46204	317-226-7290

Region or Area	Address	Phone Number
Cincinnati Area Office	36 Triangle Park Drive Building 36 Cincinnati, OH 45246	513-841-4132
Appleton Area Office	2618 North Ballard Road Appleton, WI 54915	414-734-4521
Milwaukee Area Office	310 W. Wisconsin Avenue Suite 1180 Milwaukee, WI 53203	414-297-3315
Minneapolis Area Office	110 South 4th Street Room 116 Minneapolis, MN 55401	612-348-1994
Toledo Area Office	234 N. Summit Street Room 734 Toledo, OH 43604	Lansing Area Office 801 S. Waverly Road Suite 306 Lansing, MI 48917 517-377-1892
Madison Area Office	4802 E. Broadway Madison, WI 53716	608-264-5388
Peoria Area Office	2918 Willow Knolls Road Peoria, IL 61614	309-671-7033
Eau Claire District Office	500 Barstow Street Room B-9 Eau Claire, WI 54701	715-832-9019
Fairview Hgts District Office	11 Executive Drive Suite 11 Belleville, IL 62208	618-632-8612
Region VI (Arkansas, Louisiana, New Mexico, Oklahoma, and Texas)		
Dallas Regional Office	525 Griffin Street Room 602 Dallas, Texas 75202	214-767-4731
Dallas Area Office	8344 East R.L. Thornton Freeway Suite 420 Dallas, Texas 75228	214-320-2400

Region or Area	Address	Phone Number
Austin Area Office	303 Grant Building 611 E. 6th Street Room 303 Austin, Texas 78701	512-482-5783
Albuquerque Area Office	Western Bank Building 505 Marquette, N.W. Suite 820 Albuquerque, NM 87102	505-766-3411
Baton Rouge Area Office	2156 Woodale Blvd. Hoover Annex Suite 200 Baton Rouge, LA 70806	504-389-0474
Corpus Christi Area Office	Government Plaza 400 Mann Street Room 300 Corpus Christi, TX 78401	Lubbock Area Office 1205 Texas Avenue Room 422 Lubbock, Texas 79401
Houston North Area Office	350 N. Sam Houston Pkwy, East Suite 120 Houston, Texas 77060	713-591-2438
Fort Worth Area Office	North Starr II Suite 430 8713 Airport Freeway Fort Worth, TX 76180-7604	817-885-7025
Houston South Area Office	1765 1 El Camino Real Suite 400 Houston, TX 77058	713-286-0583
Oklahoma City Area Office	420 W. Main Suite 300 Oklahoma City, OK 73102	405-231-5351
Little Rock Area Office	TCBY Building 425 West Capitol Avenue Suite 450 Little Rock, AK 72201	501-324-6291

Region or Area	Address	Phone Number
Region VII (Iowa, Kansas, Missouri, and Nebraska)		
Kansas City Regional Office	911 Walnut Street Room 406 Kansas City, MO 64106	816-426-5861
Kansas City Area Office	6200 Connecticut Avenue Room 303 Kansas City, MO 64120	816-426-2756
Des Moines Area Office	210 Walnut Street Room 815 Des Moines, IA 50309	515-284-4794
Omaha Area Office	Overland - Wolf Bldg Room 100 6910 Pacific Street Omaha, NE 68106	402-221-3182
St. Louis Area Office	911 Washington Street Room 420 St. Louis, MO 63101	314-425-4249
Wichita Area Office	301 N. Maine 300 Epic Center Wichita, KS 67202	316-269-6644
Mission District Office	5799 Broadmoor Suite 338 Mission, KS 66202	913-236-2681
Region VIII (Colorado, Montana, North Dakota, South Dakota, Utah, and Wyoming)		
Denver Regional Office	Federal Building Room 1576 1961 Stout Street Denver, CO 80204	303-844-3061
Billings Area Office	19 North 25th Street Billings, MT 59101	406-657-6649
Bismarck Area Office	Federal Building Room 348 P.O. Box 2439 Bismarck, ND 58501	701-250-4521

APPENDIX B. REGIONAL AND AREA OSHA OFFICES (Continued)

Region or Area	Address	Phone Number
Englewood Area Office	7935 E. Prentice Avenue Suite 209 Englewood, CO 80111	303-843-4500
Denver Area Office	Colonnade Center Suite 360 1391 Spear Blvd. Suite 210 Denver, CO 80204-5285	303-844-5285
Salt Lake City Area Office	1781 S. 300 West P-0. Box 15200 Salt Lake City, UT 84115	801-486-8405
Region IX (American Samoa, Arizona, California, Guam, Hawaii, Nevada, and Trust Territory of the Pacific Islands)		
San Francisco Regional Office	71 Stevenson Street Suite 420 San Francisco, CA 94105	415-744-6670
Sacramento Area Office	105 El Camino Plaza 1st Floor Sacramento, CA 95815	916-978-5641
Honolulu Area Office	300 Ala Moana Blvd Suite 5122 Honolulu, HI 96850	808-541-2685
Phoenix Area Office	3221 N. 16th Street Suite 100 Phoenix, AZ 85016	602-640-2007
San Diego District Office	5675 Ruffin Road Suite 330 San Diego, CA 92123	619-557-2909
San Francisco Area Office	71 Stevenson Street Suite 415 San Francisco, CA 94105	415-744-7120
Carson City Area Office	1050 E. Williams Street Suite 435 Carson City, NV 89701	702-885-6963

Region or Area	Address	Phone Number
Region X (Alaska, Idaho, Oregon, and Washington)		
Seattle Regional Office	111 Third Avenue Suite 715 Seattle, WA 98101-3212	206-553-5930
Anchorage Area Office	301 W. Northern Lights Blvd. Suite 407 Anchorage, AK 99503	907-271-5152
Bellevue Area Office	121 - 107th Avenue, NE Room 110 Bellevue, WA 98004	206-553-7520
Boise Area Office	3050 N. Lakeharbor Lane Suite 134 Boise, ID 83724	208-334-1867
Portland Area Office	1220 SW 3rd Avenue Room 640 Portland, OR 97204	503-326-225 1

APPENDIX C

SOURCES FOR STATE SAFETY AND HEALTH STANDARDS

APPENDIX C. SOURCES FOR STATE SAFETY AND HEALTH STANDARDS

There are 25 states and territories which operate state OSHA programs provided for under Section 18 of the Occupational Safety and Health Act of 1970. The law declares that "any state which, at any time, desires to assume responsibility for development and enforcement therein of occupational safety and health standards relating to any occupational safety and health issue with respect to which a Federal standard has been promulgated under section 6 shall submit a State plan for the development of such standards and their enforcement." State standards and their enforcement must be at least as effective in providing safe and healthful working conditions as the federal program.

Presently, 21 states and two territories (Puerto Rico and the Virgin Islands) operate complete programs covering both private and public sector employers and employees. Two states, New York and Connecticut, operate state programs covering only public sector employers and employees. Federal OSHA provides coverage of private sector employers and employees in 29 states and the District of Columbia. The Act gives federal OSHA no authority to cover public employees.

State-Plan States

State/Territory	Station Organization	Phone Number
Alaska	Alaska OSH 3301 Eagle Street Anchorage, AK 99510	(907) 264-2597
Arizona	State of Arizona OSH P.O. Box 19070 Phoenix, AZ 85005-9070	(602) 542-5795
California	CAL/OSHA 455 Golden Gate Avenue Room 5202 San Francisco, CA 94102	(415) 703-4341
Connecticut*	Connecticut Department of Labor Occupational Safety and Health Division 200 Folly Brook Blvd. Wethersfield, CT 06109	(203) 566-4500
Hawaii	State of Hawaii Department of Labor and Industrial Relations Div. of Occupational Safety & Health 830 Punchbowl Street Honolulu, HI 96813	(808) 586-0116

*Connecticut and New York have public sector programs only.

APPENDIX C. SOURCES FOR STATE SAFETY AND HEALTH STANDARDS
(Continued)

State/Territory	Station Organization	Phone Number
Indiana	Indiana Department of Labor Room 1013, State Office Building Indianapolis, IN 46204	(3 17) 232-2685
Iowa	Iowa Division of Labor 100 East Grand Des Moines, IA 50319	(515) 281-3606
Kentucky	Kentucky Occupation Safety and Health Program Kentucky Labor Cabinet 1047 U.S. 127 South, Suite 4 Frankfort, KY 40601	(502) 564-2300
Maryland	State of Maryland-MOSH Division of Labor and Industry 501 St. Paul Place Baltimore, MD 21202	(410) 333-4195
Michigan	Michigan Department of Public Health Division of Occupational Health P.O. Box 30035 Lansing, MI 48909	(517) 373-9600
Minnesota	Minnesota Occupational Health and Safety 443 Lafayette Road St. Paul, MN 55155	(612) 296-2116
Nevada	Nevada Occupational Safety and Health Enforcement Section 1370 South Curry Street Carson City, NV 89710	(702) 687-5240
New Mexico	New Mexico Occupational Health and Safety Bureau P.O. Box 26110 Santa Fe, NM 87502	(505) 827-2877

*Connecticut and New York have public sector programs only.

State/Territory	Station Organization	Phone Number
New York*	New York State Department of Labor Public Employees Safety and Health Program Room 457, Building 12 State Office Building Campus Albany, NY 12240	(518) 457-1263
North Carolina	North Carolina Department of Labor Division of Occupational Safety and Health 319 Chapanoke Road Raleigh, NC 27603-3432	(919) 662-4575
Oregon	Occupational Safety and Health Division (OR-OSHA) Labor and Industries Bldg., Room 430 Salem, OR 97310	(503) 378-3272
Puerto Rico	Department of Labor and Human Resources Occupational Safety and Health Offices 505 Munoz Rivera Avenue Hato Rey, PR 00918	(809) 754-2171
South Carolina	South Carolina Department of Labor Box 11329 Columbia, SC 29211	(803) 734-9600
Tennessee	Tennessee Department of Labor Division of Occupation Safety and Health 501 Union Building Nashville, TN 37219	(615) 741-2793
Utah	Occupational Safety and Health Division Utah Industrial Commission 160 East 300 South, Third Floor Salt Lake City, UT 84114-6650	(801) 530-6901
Vermont	Vermont Occupational Safety and Health Administration National Life Building, Drawer 20 Montepelier, VT 05602-3401	(802) 828-2765

*Connecticut and New York have public sector programs only.

State/Territory	Station Organization	Phone Number
Virgin Islands	Department of Labor 2131 Hospital Street Christiansted, St. Croix, VI 00828-4660	(809) 773-1994
Virginia	Commonwealth of Virginia Department of Labor and Industry Powers-Taylor Building 13 South Thirteenth Street Richmond, VA 23219	(804) 786-5873
Washington	Washington State Department of Labor and Industries P.O. Box 4401 Olympia, WA 98504-4001	(360) 902-4200
Wyoming	Wyoming Occupational Health and Safety Herschler Building, Second Floor, East 122 West 25th Street Cheyenne, WY 82002	(307) 777-7786

*Connecticut and New York have public sector programs only.

Appendix D

LIST OF ORGANIZATIONS WITH STANDARDS AND INFORMATION RELATED TO STATION CONSTRUCTION

APPENDIX D. LIST OF ORGANIZATIONS WITH STANDARDS AND INFORMATION RELATED TO STATION CONSTRUCTION

Organization	Address	Phone/Fax Numbers	Relevant Services
AABC Associated Air Balance Council	1518 K Street NW Washington, DC 20005	(202) 737-0202 FAX: (202) 638-4833	Ventilation and exhaust system design requirements.
AACE American Association of Cost Engineers	P.O. Box 1557 Morgantown, WV 26507	(304)296-8444 FAX:(304)291-5728	Guidelines and standards for preparing cost estimates for construction projects.
AAMA American Architectural Manufacturers Association	1540 E. Dundee Road, Suite 310 Palatine, IL 60067	(708)-202-1350 FAX: (708)202-1480	Specifications and standards for various building products. Building products information regarding strength of materials and fire resistivity .
AAMI Association for the Advancement of Medical Instrumentation	3330 Washington Blvd. Suite 400 Arlington, VA 22201	(800) 332-2264 FAX: (703) 276-0793	Standards for the selection, use and maintenance of emergency medical equipment that is used in hospitals and the fire service.
AASHTO American Association of State Highway & Transportation Officials	444 North Capital Street NW, Suite 249 Washington, DC 20001	(202) 624-5 800 FAX: (202) 624-5806	Standards for road and highway design. Refer to for specifying access road design and traffic control for fire and emergency response facilities.
ABMA American Boiler Manufacturers Association	950 North Glebe Road, Suite 160 Arlington, VA 22203-1824	(703) 522-7350 FAX: (703) 522-2665	Standards and specifications for heating and hot water systems. High efficiency equipment designs for energy conservation.

APPENDIX D. LIST OF ORGANIZATIONS WITH STANDARDS AND INFORMATION RELATED TO STATION CONSTRUCTION (Continued)

Organization	Address	Phone/Fax Numbers	Relevant Services
ABYC American Boat and Yacht Council Inc.	3069 Solomon's Island Road Edgewater, MD 21037-1416	(410) 956-1050 FAX: (410) 956-2737	Standards and specifications for the design and build of emergency response vessels such as fire boats and search and rescue equipment.
ACGIH American Conference of Governmental Industrial Hygienists Inc.	1330 Kemper Meadow Drive Cincinnati, OH 45240	(513) 742-2020 FAX: (513) 742-3355	National standards for employee safety, health and toxicological issues. Resource for ventilation and indoor air quality, hazardous materials communications, bloodborne pathogens, noise and hearing conservation, and ergonomics.
ACPA American Concrete Pavement Association	3800 North Wilke Road, Suite 490 Arlington Heights, IL 60004	(708) 394-5577 FAX: (708) 394-5610	Specification for the design and installation of concrete roads, driveways and foundations.
ACI American Concrete Institute	22400 West Seven Mile Road Detroit, MI 48219-1849	(313) 532-2600 FAX: (313) 533-4747	Testing and evaluation of structural concrete mixtures. Specifications for the selection of concrete for strength, fire resistivity, seismic integrity.

APPENDIX F. LIST OF ORGANIZATIONS WITH STANDARDS AND INFORMATION RELATED TO STATION CONSTRUCTION (Continued)

Organization	Address	Phone/Fax Numbers	Relevant Services
ACS American Chemical Society	1155 16th Street NW Washington, DC 20036	(202) 872-4600 FAX: (202) 872-4615	Standards and information for the development, use and disposal of industrial chemicals. Would be a valuable resource in selection of building coatings and determining risks of adjacent properties. Also pesticides and environmental health and safety.
ACSM American Congress on Surveying and Mapping	5410 Grosvenor Lane Suite 100 Bethesda, MD 20814	(301) 493-0200 FAX: (301) 493-8245	Standards for the surveying and mapping project sites, environmental issues, geological topography, cartography.
A E I C Association of Edison Illuminating Companies	600 North 18th. Street P.O. Box 2641 Birmingham, AL 35291-0992	(205) 250-2530 FAX: (205) 250-2540	Standards for power generation, lighting, heat and other uses of electrical power.
AEMA Asphalt Emulsion Manufacturers Association	#3 Church Circle, Suite 250 Annapolis, MD 21401	(410) 267-0023 FAX: (410) 267-7546	Standards for roadway, driveway, foundation and roof sealing products. Beneficial in the reduction of damage due to flooding, freezing and chemical agents.

APPENDIX D. LIST OF ORGANIZATIONS WITH STANDARDS AND INFORMATION RELATED TO STATION CONSTRUCTION (Continued)

Organization	Address	Phone/Fax Numbers	Relevant Services
A.G.A. American Gas Association	8501 East Pleasant Valley Road Cleveland, OH 44131	(216) 524-4990 FAX: (216) 624-3463	Testing and certification of natural gas, propane and other hydrocarbon gas fired equipment. Boilers, furnaces and dryers are influence by this organization.
AHA American Heart Association	1280 South Parker Road Denver, CO 80231	(303) 369-5433 FAX: (303) 369-8087	Guidelines for minimizing the risk of heart disease. Guidelines for establishing exercise programs and facilities, emergency services, CPR, and walk-in services.
AHAM Association of Home Appliance Manufacturers	20 North Wacker Drive Chicago, IL 60606	(312) 984-5800 FAX: (312) 984-5823	Certification or cooking and other household equipment that would be used in facilities housing employees or volunteer emergency personnel.
AHBA American Hardboard Association	1210 West NW Hwy. Palatine, IL 60067	(708) 934-8800 FAX: (708) 934-8803	Standards for the manufacture and use of wood products as a building material. Testing for strength, fire resistivity, and preservative treatments.

APPENDIX D. LIST OF ORGANIZATIONS WITH STANDARDS AND INFORMATION RELATED TO STATION CONSTRUCTION (Continued)

Organization	Address	Phone/Fax Numbers	Relevant Services
AIAA American Institute of Aeronautics and Astronautics	370 L'Enfant Promenade SW Washington, DC 20024	(202) 646-7400 FAX: (202) 646-7508	Standards and specification for the design and construction of aviation products and aircraft. Refer to for the procurement of rescue, reconnaissance, and fire fighting airplanes and helicopters.
AIA/NA Asbestos Information Association/North America	1745 Jefferson Davis Highway Crystal Square 4, Suite 509 Arlington, VA 22202	(703) 979-1150 FAX: (703) 979-1152	Guidelines for the use and remediation of asbestos products and building materials.
AIA American Institute of Architects	1735 New York Avenue NW Washington, DC 20006	(202) 626-7300 FAX: (202) 626-7420	Standards for the design and building of facilities and structures. Important resource for new construction and renovating historical buildings.
AICHE American Institute of Chemical Engineers	345 East 47th. Street New York, NY 10017	(212) 705-7657 FAX: (212) 752-3294	Standards and technical manuals for the production and use of chemical products.
AIHA American Industrial Hygiere Association	2700 Prosperity Avenue, Suite 250 Fairfax, VA 22031-4311	(703) 849-8888 FAX: (703) 207-3561	Guidelines for employee safety, health and toxicological issues. Resource for ventilation and indoor air quality, hazardous materials communications, bloodborne pathogens, noise and hearing conservation, and ergonomics.

APPENDIX D. LIST OF ORGANIZATIONS WITH STANDARDS AND INFORMATION RELATED TO STATION CONSTRUCTION (Continued)

Organization	Address	Phone/Fax Numbers	Relevant Services
AISC American Institute of Steel Construction	1 East Wacker Drive, Suite 3100 Chicago, IL 60601-2001	(3 12) 670-2400 FAX: (312) 670-5403	Standards, specifications for the fabrication and use of steel products. Design standards for steel construction.
AISE Association of Iron & Steel Engineers	Three Gateway Center, Suite 2350 Pittsburgh, PA 15222	(412) 281-6323 FAX: (412) 281-4657	Society for the design and use of steel in construction and equipment.
AISG American Insurance Services Group Inc.	85 John Street New York, NY 10038	(2 12) 669-0400 FAX: (212) 669-0535	Insurance surveys of various properties to determine risk and potential loss due to fire, wind, flood and other disasters. Valuable information for site selection or abandonment.
AISI American Iron & Steel Institute	1101 17th. Street NW, Suite 1300 Washington, DC 20036	(202) 452-7100 FAX: (202) 463-6573	Design standards and documents for the use of steel and iron products in the construction industry.
AITC American Institute of Timber Construction	7012 Revere Parkway, Suite 140 Englewood, CO 80012	(303) 792-9559 (303) 792-0669	Technical information on the design and use of timber products in the construction industry. Standards for structural integrity, fire resistivity, and preservation.

APPENDIX D. LIST OF ORGANIZATIONS WITH STANDARDS AND INFORMATION RELATED TO STATION CONSTRUCTION (Continued)

Organization	Address	Phone/Fax Numbers	Relevant Services
ALA American Lighting Association	435 North Michigan Ave., Suite 1717 Chicago, IL 60611-4067	(3 12) 644-0828 FAX: (312) 644-8557	Guidelines and standards for the manufacturing and installation of lighting equipment. Useful for the design and installation of general, emergency and security lighting systems.
ALSC American Lumber Standard Committee	P.O. Box 210 Germantown, MD 20875-0210	(301) 972-1700 FAX: (301) 540-8004	Standards for the manufacture and use of lumber and wood products for use in the construction industry.
AMCA Air Movement & Control Association Inc.	30 West University Drive Arlington Heights, IL 60004-1893	(708) 394-0150 FAX: (708) 253-0080	Standards for the application and certification of ventilation and exhaust systems. Guidelines for indoor air quality and vehicle exhaust control methods.
ANSI American National Standards Institute	11 West 42nd. Street, 13th. Floor New York, NY 10036	(212) 642-4900 FAX: (212) 398-0023	National consensus safety standards for design, installation and use of various types of equipment such as elevators, industrial trucks, machine guards. Recognized as one of the major safety standards in the United States.

APPENDIX D. LIST OF ORGANIZATIONS WITH STANDARDS AND INFORMATION
RELATED TO STATION CONSTRUCTION (Continued)

Organization	Address	Phone/Fax Numbers	Relevant Services
APA American Plywood Association	7011 South 19th. Street P.O. Box 11700 Tacoma, WA 98411-0700	(206) 565-6600 FAX: (206) 565-7265	Standards for the manufacture and use of plywood products. Certification of products by structural strength, fire resistance, and resistance to wear.
APHA American Public Health Association	1015 15th. Street NW Washington, DC 20005	(202) 789-5600 FAX: (202) 789-5661	Standards for minimizing health risk when working with public. Valuable information for providers of emergency medical services.
API American Petroleum Institute	1220 L Street NW Washington, DC 20005	(202) 682-8000 FAX: (202) 962-4776	Standards for fuel processing, storage and handling. Fire Protection standards for fuel dispensing and management.
APWA American Public Works Association	106 West 11th. Street Kansas City, MO 64105-1806	(816) 472-6100 FAX: 472-1610	Standard for public and private water supplies and sewer systems. Quality control and certification of water supply pipes, valves and other appurtenances.
AREA American Railway Engineering Association	50 F Street NW, Suite 7702 Washington, DC 20001	(202) 639-2190 FAX: (202) 639-2183	Standards for design, layout and maintenance of rail systems, right of ways, equipment.

APPENDIX D. LIST OF ORGANIZATIONS WITH STANDARDS AND INFORMATION RELATED TO STATION CONSTRUCTION (Continued)

Organization	Address	Phone/Fax Numbers	Relevant Services
ARI Air-Conditioning & Refrigeration Institute	4301 North Fairfax Drive, Suite 425 Arlington, VA 22203	(703) 524-8800 FAX: (703) 528-3816	Standards and guidelines for air-conditioning and ventilation systems.
ARM Asphalt Roofing Manufacturers Association	6001 Executive Blvd., Suite 201 Rockville, MD 20852	(301) 231-9050 FAX: (301) 881-6572	Quality control standards and product certification of roofing systems and materials.
ARTBA American Road & Transportation Builders Association	1010 Massachusetts Ave. NW Washington, DC 20001	(202) 289-4434 FAX: (202) 289-4435	Standards the development and construction of roads and rail systems.
ASCE American Society of Civil Engineers	345 East 47th. Street New York, NY 10017	(212) 705-7510 FAX: (212) 705-7712	Standards, manuals, reports on engineering practice relating to structures, concrete, steel and design practices.
ASHRAE American Society of Heating, Refrigerating & Air-Conditioning Engineers Inc.	1791 Tullie Circle NE Atlanta, GA 30329-2305	(404) 636-8400 FAX: (404) 321-5478	Standards and certification of heating, cooling, exhaust and ventilation systems and equipment.
ASME American Society of Mechanical Engineers	345 East 47th. Street New York, NY 10017	(212) 705-7722 FAX: (212) 753-9568	Standards, certification of boilers, pressure vessels, welding, and other mechanical systems.

APPENDIX D. LIST OF ORGANIZATIONS WITH STANDARDS AND INFORMATION RELATED TO STATION CONSTRUCTION (Continued)

Organization	Address	Phone/Fax Numbers	Relevant Services
ASNT American Society for Nondestructive Testing Inc.	1711 Arlington Lane P.O. Box 28515 Columbus, OH 43228-0518	(800) 222-2768 FAX: (614) 274-6899	Testing standards for metal fatigue, concrete strength and other material assemblies. Listing of qualified testing agencies and vendors.
ASQC American Society for Quality Control	611 East Wisconsin Avenue Milwaukee, WI 53202-4606	(414) 272-8575 FAX: (414) 272-1734	Standards and guidelines for quality control and assurance.
ASSE American Society of Safety Engineers	1800 East Oakton Des Plaines, IL 60018-2187	(708) 692-4121 FAX: (708) 296-3769	Society of engineers, technicians and consultants specializing in the field of safety engineering. Publications and interpretation of safety laws, ordinances and standards.
ASSE American Society of Sanitary Engineering	P.O. Box 40362 Bay Village, OH 44140	(216) 835-3040 FAX: (216) 835-3488	Standards and guidelines for the management of solid waste and hazardous materials.
ASTM American Society for Testing & Materials	100 Barr Harbor Drive W. Coshohocken, PA 19428-2959	(610) 832-9500 FAX: (610) 832-9555	Certification of testing and materials laboratories. Standards and guidelines on a variety of building supplies and other products.
ATA Air Transport Association of America	P.O. Box 511 Annapolis Junction, MD 20701	(202) 626-4000 FAX: (202) 862-0570	Specifications and standards for the aviation industry which can apply to departments with air operations.

APPENDIX D. LIST OF ORGANIZATIONS WITH STANDARDS AND INFORMATION
RELATED TO STATION CONSTRUCTION (Continued)

Organization	Address	Phone/Fax Numbers	Relevant Services
ATA American Trucking Association	2200 Mill Road Alexandra, VA 22314-4677	(800) 282-5463 FAX: (800) 225-8382	Standards for the trucking industry.
ATI American Textile Institute	1801 K Street, Suite 900 Washington, DC	(202) 862-0500 FAX: (202) 862-0570	Standards for the manufacture and treatment of textiles. Fire resistive treatments for finish materials.
AWCI Association of The Wall & Ceilings Industries International	307 East Annandale Road Falls Church, VA 22042-2433	(703) 534-8300 FAX: (703) 534-8307	Standards for wall and ceiling assemblies and materials.
AWI Architectural Woodwork Institute	P.O. Box 1550 Centerville, VA 22020	(703) 222-1100 FAX: (703) 222-2499	Standards for use, function and selection of woodwork products.
AWPA American Wood-Preservers' Association	1945 Old Gallows Road, Suite 550 Vienna, VA 22182	(703) 893-4005 FAX: (703) 893-8492	Standards for the preservation of wood products. Fire-resistivity, moisture and insect decay, hazard material management.
AWS American Welding Society Inc.	550 Northwest LeJeune Road P.O. Box 351040 Miami, FL 33135	(305) 443-9353 FAX: (305) 443-7559	Specifications, codes and standards for safety health and quality assurance of welding operations.

APPENDIX D. LIST OF ORGANIZATIONS WITH STANDARDS AND INFORMATION RELATED TO STATION CONSTRUCTION (Continued)

Organization	Address	Phone/Fax Numbers	Relevant Services
AWWA American Water Works Association	6666 West Quincy Avenue Denver, CO 80235	(303)794-7711 FAX: (303) 794-7310	Standard for public and private water supplies and sewer systems. Quality control and certification of water supply pipes, valves and other appurtenances.
BCI Battery Council International	401 N. Michigan Ave. Chicago, IL 60611	(312) 644-6610 FAX: (312) 321-6869	Standards for the design of electrical storage batteries, safe handling and storage. UPS and emergency power systems..
BIA Brick Institute of America	11490 Commerce Park Drive, Suite 300 Reston, VA 22091	(703) 620-0010 FAX: (703) 620-3928	Standards for architectural, engineering, paving and construction with brick. Seismic standards for brick assemblies.
BIFMA Business & Industrial Furniture Manufacturers Association	2680 Horizon Drive SE, Suite A-1 Grand Rapids, MI 49546	(616) 285-3963 FAX: (616) 285-3765	Design and manufacturing standards for furniture. Ergonomic, fire resistivity, and functional standards for specifying furniture.
BOCA Building Officials & Code Administrators International Inc.	4051 West Flossmoor Road Country Club Hills, IL 60478-5795	708) 799-2300 FAX: (708) 799-4981	Building codes and standards adopted by many jurisdictions in the East and Midwest. Occupancy classifications, structural requirements, exiting, lighting, ventilation.

APPENDIX D LIST OF ORGANIZATIONS WITH STANDARDS AND INFORMATION
RELATED TO STATION CONSTRUCTION (Continued)

Organization	Address	Phone/Fax Numbers	Relevant Services
BSSC Building Seismic Safety council	1201 L Street NW Suite 400 Washington, DC 20005	202-289-7800 FAX: 202-289-7810	Provides details reports and guidelines for rehabilation of old buildings and new construction meeting seismic requirements
CAB0 Council of American Building Officials	5203 Leesburg Pike, Suite 708 Falls Church, VA 22041	(703) 931-4533 FAX: (703) 379-1546	Standard making consortium for the development of building codes and standards primarily for the southeastern U.S.
CAGI Compressed Air & Gas Institute	1300 Summer Ave. Cleveland, OH 44155	(216) 241-7333 FAX: (216) 241-0105	Standards for compressed air and gas systems such as cylinders used in cascade systems.
CED Civil Engineering Data	Department of Army Waterways Experiment Station Corps of Engineers P.O. Box 631 Vicksburg, MS 39180-6199	(601) 634-2571 FAX: (601) 634-2452	Standards for engineers on soils and structures research. Technical papers and reports.
CEMA Conveyor Equipment Manufacturers Association	932 Hungerford Drive, Suite 36 Rockville, MD 20850	(301) 738-2448 FAX: (301) 738-0076	Standards for material handling equipment.
CGA Compressed Gas Association Inc.	1725 Jefferson Davis Hwy., Suite 1004 Arlington, VA 22202-4100	(703) 412-0900 FAX: (703) 412-0128	Standards for the storage and handling of compressed gases. Safety standards for cylinder storage, pressure relief valves and piping systems.

Organization	Address	Phone/Fax Number	Relevant Services
CISCA Ceilings & Interior Systems Contractors Association	579 West North Avenue Ehnhurst, IL 60126	(708) 833-1919 FAX: (708) 833-1940	Guidelines and develpoment of ceiling and interior finish systems.
CISPI Cast Iron Soil Pipe Institute	5959 Shallowford Road, Suite 419 Chattanooga, TN 37421	(615) 892-0137 FAX: (615) 892-0817	Standards for the manufacture and installation of cast iron piping system.
CLFMI Chain Link Fence Mfrs. Institue	1776 Massachuesetts Ave. NW, Suite 500 Washington, DC 20036	(202) 659-3537 FAX : (202) 857-1220	Standards for the design development and installation of chain link fencing.
CMA Cookware Manufacturers Association	PO Box 531335 Birmingham, AL 35253-1335	(205) 802-7600 FAX: (205) 802-7610	Industrial and commercial standards for cookware equipment.
CMAA Crane Manufacturers Association of America, Inc.	8720 Red Oak Blvd., Suite 201 Charlotte, NC 28217	(704) 522-8644 FAX: (704) 522-7826	Standards and product development for cranes and material handling systems.
CPB Contractors Pump Bureau	111 East Wisconsin Ave., Suite 940 Milwaukee, WI 53202	(414) 272-0943 FAX: (414) 272-1170	Clearinghouse for pump manufacturer's information.
CRMA Commercial Refrigerator Manufacturers Association	1101 Connecticut Ave. NW, Suite 700 Washington, DC 20036	(202) 857-1145 FAX: (202) 223-4579	Information for commercial and residential refrigeration equipment.
CRSI Concrete Reinforcing Steel Institute	933 North Plum Grove Road Schaumburg, IL 60173-4758	(708) 517-1200 FAX: (708) 517-1206	Guidelines, standards and testing information pertaining to structural concrete assemblies.

APPENDIX D. LIST OF ORGANIZATIONS WITH STANDARDS AND INFORMATION RELATED TO STATION CONSTRUCTION (Continued)

Organization	Address	Phone/Fax Numbers	Relevant Services
CSDA Concrete Sawing & Drilling Association	4900 Blazer Parkway Dublin, OH 43017	(614) 766-3656 FAX: (614) 766-3605	Standards and equipment information regarding concrete sawing and drilling.
CSI Construction Specifications Institute Inc.	601 Madison Street Alexandra, VA 22314	(703) 684-0300 FAX: (703) 684-0465	Standard specifications for construction project design documents.
DEMA Diesel Engine Manufacturers Association	30200 Detroit Road Cleveland, OH 44145-1967	(216) 899-0010 FAX: (216) 892-1404	Standards and product testing information. Fire pump and electric generator drive motors.
DHI Door & Hardware Institute	14170 Newbrook Drive Chantilly, VA 22021-2223	(703) 222-2010 FAX: (703) 222-2410	Standards and specifications for door hardware. Fire ratings, ADA compliance, product test information.
EGSA Electrical Generating Systems Association	10251 B West Sample Road PO Box 9257 Coral Springs, FL 33065	(305) 755-2677 FAX: (305) 755-2679	Information regarding standard and emergency power generation.
EIMA Exterior Insulation Manufacturers Association	2759 State Road 580, Suite 112 Clearwater, FL 34621	(813) 726-6477 FAX: (813) 726-8180	Standard for insulation as a building material.
EJMA Expansion Joint Manufacturers Association	25 North Broadway Tarrytown, NY 10591	(914) 332-0040 FAX: (914) 332-1541	Standards and information for expansion joint products. Useful in construction for seismic and structural strain relief.

APPENDIX D. LIST OF ORGANIZATIONS WITH STANDARDS AND INFORMATION
RELATED TO STATION CONSTRUCTION (Continued)

Organization	Address	Phone/Fax Numbers	Relevant Services
ESD Electrical Overstress/Electrostatic Discharge Association Inc.	200 Liberty Plaza Rome, NY 13440	(315) 339-6937 FAX: (315) 339-6793	Standards for static electricity and lightning protection.
FEMA/USFA Federal Emergency Management Agency/ U. S. Fire Administration	16825 S. Seton Avenue Emmitsburg, MD 21727	(301) 447-1000 FAX: (301) 447-1213	Guidelines for station construction, fire service safety and health publications.
FGMA Flat Glass Marketing Association	White Lakes Professional Bldg. 3310 Harrison Street Topeka, KS 66611-2279	(913) 266-7013 FAX: (913) 266-0272	Standards for manufacturing and installation of glass products. Fire rated and tempered glass product information.
FGCC Federal Geodetic Control Committee	1315 East West Highway Silver Spring, MD 20910	(301) 713-3242 FAX: (301) 713-4172	Information on seismic conditions, tidal waves, flood plains and other potential natural threats.
FHWA Federal Highway Administration	Office of Highway Safety 400 Seventh Street SW Washington, DC 20590-0001	(202) 366-0411 FAX: (202) 366-8518	Highway safety standards. Roadway construction, traffic control and road maintenance.
FMERC Factory Mutual Engineering & Research Corp.	1151 Providence Highway PO Box 9102 Norwood, MA 02062	(617) 762-4300 FAX: (617) 762-9375	Independent testing laboratory for the fire research on many manufactured products. Fire sprinkler systems, roof assemblies, wall assemblies and approved products guide.

APPENDIX D. LIST OF ORGANIZATIONS WITH STANDARDS AND INFORMATION RELATED TO STATION CONSTRUCTION (Continued)

Organization	Address	Phone/Fax Numbers	Relevant Services
GA Gypsum Association	810 First Street NE #510 Washington, DC 20002	(202) 289-5440 FAX: (202) 289-3707	Standards for the use of gypsum products. Fire rating and structural test information.
GAMA General Aviation Manufacturers Association	1400 K Street NW, Suite 801 Washington, DC 20005	(202) 393-1500 FAX: (202) 842-4063	Standards and product information. Useful for fire and rescue ah-fleets.
HEI Heat Exchange Institute	1300 Sumner Ave. Cleveland, OH 44115-2180	(216) 241-7333 FAX: (216) 241-0105	Standards and information for heating and cooling equipment.
HFES Human Factors and Ergonomics Society	PO Box 1369 Santa Monica, CA 90406	(310) 394-1811 FAX: (310) 394-2410	Standards and information for the study and prevention of injuries caused by poorly designed work spaces; ergonomic guidelines for work space design.
HPMA Hardwood Plywood & Veneer Association	1825 Michael Farady Drive PO Box 2789 Reston, VA 22090-2789	(703) 435-2900 FAX: (703) 435-2537	Standards and product information on hardwood floors.
HVA Air Movement & Control Association/Home Ventilation Institute (Division 30)	West University Drive Arlington Heights, IL 60004	(708) 394-0150 FAX: (708) 253-0088	Standards and information on ventilation and indoor air quality.
IAFC Internnational Association cf Fire Chiefs	4025 Fair Ridge Road Fairfax, VA 22033	(703) 273-0911 FAX: (703) 273-9363	Policies and guidelines for station design.

APPENDIX D. LIST OF ORGANIZATIONS WITH STANDARDS AND INFORMATION RELATED TO STATION CONSTRUCTION (Continued)

Organization	Address	Phone/Fax Numbers	Relevant Services
IAFF International Association of Fire Fighters	1750 New York Ave., N.W. Washington, DC 20006	(202) 737-8484 FAX: (202) 737-8418	Publications on fire fighter safety; policies and guidelines for station design.
IAPMO International Association of Plumbing & Mechanical Officials	20001 Walnut Drive South Walnut, CA 91789-2825	(714) 595-8449 FAX: (714) 594-3690	Standards, materials, installation and product materials on plumbing fixtures and piping.
ICAC Institute of Clean Air Companies	1707 L Street NW, Suite 570 Washington, DC 20036	(202) 457-0911 FAX: (202) 331-1388	Standards and information on electrostatic, fabric and charcoal air filtering systems.
ICBO International Conference of Building Officials	5360 S. Workman Mill Rd. Whittier, CA 90601	(3 10) 699-0541 FAX: (310) 692- 3853	Uniform Building Codes, publications, technical references. Building code that is used by many municipalities in the Western U.S.
IEEE Institute of Electrical & Electronics Engineers	445 Hoes Lane PO Box 1331 Piscataway, NJ 08855-1331	(800) 678-4333 FAX: (908) 562-1571	Standards for the electrical and electronics industry.
I.E.S. Institute of Environmental Sciences	940 East Northwest Hwy. Mount Prospect, IL 60056	(708) 255-1561 FAX: (708) 255-1699	Technical publications, contamination control documents.
IESNA Illuminating Engineering Society of North America	120 Wall Street, 17th. Floor New York, NY 10005	(212) 248-5000 FAX: (212) 248-5017	Handbook, lighting energy management, design guides and technical memos.

Organization	Address	Phone/Fax Numbers	Relevant Services
IMAC International Mobile Air Conditioning Association	PO Box 9000 Fort Worth, TX 76147-2000	(817) 338-1100 FAX: (817) 338-1451	Standards and products for the mobile air conditioning systems.
IME Institute of Makers of Explosives	1120 19th. Street NW, Suite 310 Washington, D.C. 20036-3605	(202) 429-9280 FAX: (202) 293-2420	Safety library publications.
IMSA International Municipal Signal Association	165 East Union Street PO Box 539 Newark, NY 14513	(315) 331-2182 FAX: (315) 331-8205	Information on design of traffic control systems.
IRI Industrial Risk Insurers	85 Woodland Street Hartford, CT 06102	(203) 520-7300 FAX: (203) 549-5780	Loss control and risk management standards.
ISDSI Insulated Steel Door Systems Institute	30200 Detroit Road Cleveland, OH 44145-1967	(216) 899-0010 FAX: (216) 892-1404	Standards, fire test data and product information.
ISS Iron & Steel Society	410 Commonwealth Drive Warrendale, PA 15086	(412) 776-9460 FAX: (412) 776-0430	Information on steel door products such as for apparatus bay doors.
ITE Institute of Transportation Engineers	525 School Street SW, Suite 410 Washington, D.C. 20024	(202) 554-8050 FAX: (202) 863-5486	Safety standards for transportation systems.
KCMA Kitchen Cabinet Makers Association	1899 Preston White Drive Reston, VA 22091-4326	(703) 264-1690 FAX: (703) 620-6530	Standards and products for kitchens and cafeterias.

APPENDIX D. LIST OF ORGANIZATIONS WITH STANDARDS AND INFORMATION RELATED TO STATION CONSTRUCTION (Continued)

Organization	Address	Phone/Fax Numbers	Relevant Services
LPI Lightning Protection Institute	3365 North Arlington Heights Rd., Suite J Arlington Heights, IL 60004	(800) 488-6864 (708) 255-3003 FAX: (708) 577-7276	Safety standards for reducing risk of lightning hazards.
LSGA Laminators Safety Glass Association	White Lakes Professional Bldg. 3310 SW Harrison Street Topeka, KS 66611-2279	(913) 266-7013 FAX: (913) 266-0272	Safety standards and product information on reinforced safety glass.
MBMA Metal Building Manufacturers Association	1300 Sumner Ave. Cleveland, OH 44115-2180	(216) 241-7333 FAX: (216) 241-0105	Standards and product information on metal building construction techniques.
MDA Marking Device Association	435 N. Michigan Ave., Suite 1717 Chicago, IL 60611	(312) 644-0828 FAX: (312) 644-8557	Standards and product information for signage.
MHI Material Handling Institute Inc.	8720 Red Oak Blvd., Suite 201 Charlotte, NC 28217	(800) 345-1815 FAX: (704) 522-7826	Standards and product information.
MICA Midwest Insulation Contractors Association	2017 South 139th. Cicle Omaha, NE	(402) 342-3463 FAX: (402) 330-9702	Guidelines for installation of thermal insulation in commercial structures
MLA Metal Lath/Steel Framing Association	600 South Federal Street, Suite 400 Chicago, IL 60605-1842	(312) 922-6222 FAX: (312) 922-2734	Standards and product information on metal lathe in wall construction.

APPENDIX D LIST OF ORGANIZATIONS WITH STANDARDS AND INFORMATION RELATED TO STATION CONSTRUCTION (Continued)

Organization	Address	Phone/Fax Numbers	Relevant Services
MMSA Materials & Methods Standards Association	614 Monroe Street Grand Haven, MI 49417	(616) 842-7844 FAX: (616) 842-1547	Materials and methods standards bulletins and current information.
MSS Manufacturers Standardization Society of the Valve & Fittings Industry	127 Park Street NE Vienna, VA 22180	(703) 281-6613 FAX: 703-281-6671	Standards and product information on construction hardware.
NAAMM National Association of Architectural Metal Manufacturers	11 South Lasalle St., Suite 1400 Chicago, IL 60603	(312) 201-0101 FAX: (312) 201-0214	Standards and product information.
NAGDM National Association of Garage Door Manufacturers	1300 Sumner Ave. Cleveland, OH 44115-2851	(216) 241-7333 FAX: (216) 241-0105	Product information on apparatus doors.
NAPHCC National Association of Plumbing-Heating-Cooling Contractors	PO Box 6808 Falls Church, VA 22046	(703) 237-8100 FAX: (703) 237-7442	Listing of qualified contractors for heating, cooling, and plumbing construction.
NCMA National Concrete Masonry Association	2302 Horse Pen Road PO Box 781 Herndon, VA 22071	(703) 713-1900 FAX: (703) 435-1910	Standards and product information on concrete foundations, walls, and other structures.
NCS National Conference of States on Building Codes and Standards Inc.	505 Huntmar Park Drive, Suite 210 Herndon, VA 22020	(703) 437-0100 FAX: (703) 481-3596	Recommended national building codes.

APPENDIX D. LIST OF ORGANIZATIONS WITH STANDARDS AND INFORMATION RELATED TO STATION CONSTRUCTION (Continued)

Organization	Address	Phone/Fax Numbers	Relevant Services
NECA National Electrical Contractors Association	3 Bethesda Metro Centre, Suite 1100 Bethesda, MD 20814	(301) 657-3110 FAX: (301) 961-6495	Listing of qualified contractors for electrical installation and repair.
NE11 National Elevator Industry Inc.	185 Bridge Plaza North, Room 310 Fort Lee, NJ 07024	(201) 944-3211 FAX: (201) 944-5483	Standards and information related to elevator design and installation.
NEMA National Electrical Manufacturers Association	2101 L Street NW, Suite 300 Washington D .C. 20037	(202) 457-8400 FAX: (202) 457-8411	Standards and product information on electrical wiring installation and repair.
NETA International Electrical Testing Association	PO Box 687 Morrison, CO 80465	(303) 697-8441 FAX: (303) 697-8431	Standards for testing electrical circuits and appliances.
NFPA National Fire Protection Association	One Batterymarch Park PO Box 9101 Quincy, MA 02269-9101	(617) 770-3000 FAX: (617) 770-0700	Fire protection standards and product information. General authority and reference standard on many fire and electrical protection issues. Also standards for fire department station safety and health.
NFP(A) National Fluid Power Association	3333 North Mayfair Road Milwaukee, WI 53222-3219	(414) 778-3344 FAX: (414) 778-3361	Standard practices and information for estimating building water flow requirements.
NFSA National Fire Sprinkler Association	4 Robin Hill Corporate Park P. 0. Box 1000 Patterson, NY 12563	(914) 878-4200 FAX: (914) 878-4215	Standard practices and guidelines for the selection and installation of fire sprinkler systems

APPENDIX D. LIST OF ORGANIZATIONS WITH STANDARDS AND INFORMATION RELATED TO STATION CONSTRUCTION (Continued)

Organization	Address	Phone/Fax Numbers	Relevant Services
NIAC National Insulation and Abatement Contractors Association	2017 South 139th. Circle Omaha, NE 68144-2149	(402) 342-3463 FAX: (402) 330-9702	Safety practices for proper installation of thermal insulation and abatement practices for asbestos removal.
NOCSAE National Operating Committee on Standards for Athletic Equipment	c/o National Federation of State High School Associations 11724 NW Plaza Circle Kansas City, MO 64153	(816) 464-5470 FAX: (816) 464-5571	Guidelines for selection, care, and maintenance of athletic equipment.
NPCA National Pest Control Association	8100 Oak Street Dunn Loring, VA 22027	(703) 573-8330 FAX: (703) 573-4116	Standards and product information on different types of pesticides and their application.
NPGA National Propane Gas Association	1600 Eisenhower Lane, Suite 100 Lisle, IL 60532	(708) 515-0600 FAX: (708) 515-8774	Standards and product information on use of propane.
NRCA National Roofing Contractors Association	102255 W. Higgins Road Suite 600 Rosemont, IL 60018-5607	(708) 299-9070 FAX: (708) 299-1183	Guidelines for roof materials and construction.
NSC National Safety Council	PO Box 558 Itasca, IL 60143-0558	(708) 285-1121 FAX: (708) 285-0797	Extensive library for several safety areas including ergonomics, electrical safety, indoor air quality, fall protection, and others.
NSWMA National Solid Waste Management Association	1730 Rhode Island Ave. NW, Suite 1000 Washington, D.C. 20036	(202) 659-4613 FAX: (202) 775-5917	Guidelines and recommended practices for managing solid waste disposal.

APPENDIX D. LIST OF ORGANIZATIONS WITH STANDARDS AND INFORMATION
RELATED TO STATION CONSTRUCTION (Continued)

Organization	Address	Phone/Fax Numbers	Relevant Services
NTIAC Nondestructive Testing Information Analysis Center/Texas Research Institute	415A Crystal Creek Drive Austin, TX 78746	(512) 263-2106 FAX: (512) 263-3530	Library for testing and product information on subjects related to non-destructive testing of building and other materials.
NTMA National Terrazzo & Mosaic Association Inc.	3166 Des Plaines Ave., Suite 132 Des Plaines, IL 60018	(708) 635-7744 FAX: (800) 323-9736 FAX: (708) 635-9127	Standards and information on stucco-based products.
NVFC National Volunteer Fire Council	1050 17th Street NW Suite 701A Washington, DC 20036	202-887-5700 FAX: (202) 887-5291	Policies and guidelines for station design.
OPEI Outdoor Power Equipment Institute, Inc.	341 S. Patrick Street Alexandria, VA 22314	(703) 549-7600 FAX: (703) 549-7604	Standards and product information.
PCI Precast/Prestressed Concrete Institute	175 West Jackson Blvd., Suite 1859 Chicago, IL 60604	(312) 786-0353	Standards and product information on precast concrete use in building construction.
PDI Plastic Drum Institute	1275 K Street NW Washington, DC 20005	(202) 371-5200 FAX: (202) 371-1022	Standards and product information on plastic containers for chemical and other liquid storage.
PDCA Painting & Decorating Contractors of America	3913 Old Lee Hwy., Suite 33B Fairfax, VA 22030	(703) 359-0826 FAX: (703) 359-2576	List of qualified contractors for interior decorating and space planning.

APPENDIX D. LIST OF ORGANIZATIONS WITH STANDARDS AND INFORMATION RELATED TO STATION CONSTRUCTION (Continued)

Organization	Address	Phone/Fax Number	Relevant Services
PDI Plumbing & Drainage Institute	1106 W. 77th. Street South Drive Inianapolis, IN 46260-3318	(317) 251-6970 FAX: (317) 251-6970	Standards and product information on plumbing and drain fixtures
PPI Plastic Pipe Institute	1275 K Street NW Washington, DC 20007	(202) 371-5306 FAX: (202) 342-0702	Standards and product information
PSTC Pressure Sensitive Tape Council	401 North Michigan Ave. Chicago, IL 60611-4267	(312) 644-6610 FAX: (312) 245-1085	Standards and product information. Used in overhead doors and entry ways.
PTI Post Tensioning Institute	1717 West Northern Ave., Suite 218 Phoenix, AZ 85021	(602) 870-7540 FAX: (602) 870-7541	Standards and product inforation on setting building posts
SBCCI Southern Building Code Congress International Inc.	900 Montclair Road Birmingham, AL 35213-1206	(205) 591-1853 FAX: (205) 591-0775	Standards and product code general used in the Southern US.
SDI Steel Deck Institute	PO Box 9506 Canton, IL 4471 1-9506	(216) 493-7886 FAX: (216) 493-7886	Standards and product information on steel decks and platforms.
SDI Steel Door Institute	30200 Detroit Road Cleveland, OH 44145-1967	(216) 899-0010 FAX: (216) 892-1404	Standards and product information on commercial entry and interior steel doors.
SJI Steel Joist Institute	1205 48th. Ave. N, Suite A Myrtle Beach, NC 29577	(803) 449-0487 FAX: (803) 449-1343	Standards and product information.

APPENDIX D. LIST OF ORGANIZATIONS WITH STANDARDS AND INFORMATION RELATED TO STATION CONSTRUCTION (Continued)

Organization	Address	Phone/Fax Numbers	Relevant Services
SMACNA Sheet Metal & Air Conditioning Contractors National Association Inc.	4201 Lafayette Center Drive Chantilly, VA 22021	(703) 803-2980 FAX: (703) 803-3732	List of qualified sheet metal and HVAC contractors by area.
SMF Snell Memorial Foundation	PO Box 493 St. James, NY 11780	(5 16) 862-6440 FAX: (516) 862-6545	Standards and product information relating to head protection.
SNAME Society of Naval Architects & Marine Engineers	601 Pavonia Ave. Jersey City, NJ 07306	(201) 798-4800 FAX: (201) 798-4975	Standards and product information relating to the maritime industry.
SSFI Scaffolding & Shoring & Forming Institute Inc.	1300 Sumner Ave. Cleveland, OH 44115	(216) 241-7333 FAX: (216) 241-0105	Standards and product information on erection of scaffolding for semi-permanent use.
SSPC Steel Structures Painting Council	4516 Henry Pittsburgh, PA 15213	(412) 687-1113 FAX: (412) 687-1153	Recommended practices for painting metal structures.
SSPMA Sump & Sewage Pump Manufacturers Association	PO Box 647 Northbrook, IL 60065-0647	(708) 559-9233 FAX: (708) 559-9235	Standards and product information on commercial sump pumps and their requirements.
STI Steel Tank Institute	570 Oakwood Road Lake Zurich, IL 60047	(708) 438-8265 (800) 822-3186 FAX: (708) 438- 4509	Standards and product information on specification and use of steel above and below ground tanks.

APPENDIX D. LIST OF ORGANIZATIONS WITH STANDARDS AND INFORMATION
RELATED TO STATION CONSTRUCTION (Continued)

Organization	Address	Phone/Fax Numbers	Relevant Services
SWI Steel Window Institute	1300 Sumner Ave. Cleveland, OH 44115	(216) 241-7333 FAX: (216) 241-0105	Standards and product information on steel windows.
TTMA Truck Trailer Manufacturers Association	1020 Princess Street Alexandria, VA 22314	(703) 549-3010 FAX: (703) 549-3014	Information on use of trailers as means for auxiliary storage.
UL Underwriters Laboratories Inc.	333 Pfingsten Road Northbrook, IL 60062	(708) 272-8800 FAX: (708) 272-8129	Standards and product safety information. Listed products catalogue of electrical, building assemblies, fire and security equipment.
VFIS Volunteer Firemen's Insurance Services			
WQA Water Quality Association	4151 Naperville Road Lisle, IL 60532	(708) 505-0160 FAX: (708) 505-9637	Standards and test methods for water quality.
WRI Wire Reinforcement Institute Inc.	1101 Connecticut Ave. NW, Suite 700 Washington, DC 20036	(202) 429-5125 FAX: (202) 223-4579	Standards and information on metal reinforcing products for the construction industry.
WWPA Western Wood Products Association	Yeon Bldg. 522 SW Fifth Avenue Portland, OR 97204-2122	(503) 224-3930 FAX: (503) 224-3934	Sample specifications for wood use in construction.

APPENDIX E

ANNOTATED BIBLIOGRAPHY
FIRE AND EMS STATION DESIGN
FOR SAFETY AND HEALTH

APPENDIX E. ANNOTATED BIBLIOGRAPHY ON
FIRE AND EMS STATION DESIGN fOR SAFETY AND HEALTH

Station Location/Site Planning

1. Bailey, Glen R., Task Force Concept in Determining Facilities Needs for the Fire Service," part of Executive Fire Officer Program Series, National Fire Academy, Emmitsburg, MD, August 1993.

    Study describes how one community anticipates the needs for new fire stations and the decisions that go into the process for locating stations.

2. Campbell, Cary E., "Planning a Small Community Fire Station," part of Executive Fire Officer Program, National Fire Academy, Emmitsburg, MD, November 1990.

    Describes one department's evaluation of site appropriateness for its facility between designing a new station or combining with existing facility. Provides a series of questions to be asked to determine need for a new station and gives financial considerations.

3. Dangler, William P., "Why Build a New Central Fire Station in Ypsilanti Township? What Will It Accomplish?," part of Executive Fire Officer Program, National Fire Academy, Emmitsburg, MD, May 1991.

    The report examines how the needs of a specific station are assessed in terms of the history of the department, a review of recommended stations and regulations, and interviews with department personnel.

4. Daverman Associates, "Evaluation of Alternative Locations for New Public Safety Facilities," Report for Greenville, Michigan, Daverman Associates, Washington DC, January 1981.

    Illustrates example process to deciding on fire station locations.

5. De Silva, Paul J., "From the Ground Up: Constructing or Reconstructing a Modern Firehouse" (part two of a series), Fire Engineering, January, 1990, pp 32-35.

    Site selection and station layout are discussed for new or renovated firehouses.

6. Gallagher, Jim, "From the Ground Up," Fire Command, November, 1989, pp. 33-36.

    Article addresses issues associated with planning a new fire station, from determining if a new station is needed to evaluating potential sites for obstructions, zoning problems, etc.

7.   Harmer, Thomas A., "Establishing Criteria for When to Build New Fire Stations," part of Executive Fire Officer Program Series, National Fire Academy, Emmitsburg, MD, July 1993.

   Provides example cases of Tallahassee Fire Department examination of fire station construction timing based on a needs assessment.

8.   Kersey, Lenore S. and Attahiru Sule Alfa, "Improved Distance Estimation Method for Location Modeling," Journal of Urban Planning and Development, Vol. 116, No. 2, Sept. 1990, pp. 100-105.

   Methods of determining travel time between points are discussed. An improved distance-estimation method that considers barriers is presented.

9.   McCarthy, Donald, "Advantages of Rehabilitation of Fire Stations in Older Communities," part of Executive Fire Officer Program Series, National Fire Academy, Emmitsburg, MD, November 1991.

   This study recommends rehabilitation of existing stations over new construction for many Northeastern U.S. fire departments based on community benefits and financial savings. The study found that federal funding could be sought to support rehabilitation efforts, that department services remained equal following rehabilitation, and that community benefits are realized during most rehabilitation projects.

10. Mott, Douglas S., "The Building of a Satellite Fire Station," part of Executive Fire Officer Program series, National Fire Academy, Emmitsburg, MD, March 1990.

   Discusses one department 's planning process and procedures to meet the demands of rapid community growth by building a satellite station.

11. Reilly, Jack M. and Pitu B. Mirchandani, "Development and Application of a Fire Station Placement Model," Fire Technology, Vol. 21, August 1985, pp. 181-98.

   Static and dynamic models for fire station placement are briefly reviewed. A model incorporating first and second unit response time distributions as well as other useful constraints is developed and applied to a "real life" situation.

12. Scheel, Duane, "Threshold Concept: When to Build a Fire Station," part of Executive Fire Officer Program Series, National Fire Academy, Emmitsburg, MD, November 1990.

List criteria and considerations for determining when to build afire station. This study showed that IS0 response guidelines were the major determining factors. The threshold concept was developed a tool to examine different scenarios that could develop as the city or community grew.

13. Smith, Carl L., "The Development of Criteria for Opening a Fire Station for the Aurora, Colorado, Fire Department," part of Executive Fire Officer Program Series, National Fire Academy, Emmitsburg, MD, March 18, 1994.

    Addresses the research questions: (1) what are the criteria for opening a station? (2) can different levels of fire protection be defined for both rural and urban areas and what are the different requirements for rural and urban fire protection? (3) what is the minimum staffing for both a rural and urban fire station? (4) what equipment should be housed in each type of station?, and (5) what is the optimal size of afire station in both urban and rural areas? Applies these questions to the Aurora, Colorado Fire Department.

14. Wabich, John A., "Planning the Fire Station to Meet Department and Community Needs," International Fire Chief, Vol. 46(2), February 1980, pp. 20-23.

    Provides perspectives for balancing department and community needs for locating and designing a fire station.

Finance/Management

1. Amestoy, Lee, "Fire Station Construction - An Analysis of the Importance of Architectural Services," part of Executive Fire Officer Program series, National Fire Academy, Emmitsburg, MD, December, 1991.

    Discusses the need for qualified, professional architects and cescribes a program for the budgeting, selection and establishing of a working relationship with the architect.

2. Bell, Michael P., James L. Carswell, Mike Green, Robert C. May, A. Requate, and Andrew J. Rocca, "Fiscal Impact to Fire Stations Due to Woment in the Fire Service," part of Executive Fire Officer Program Series, National Fire Academy, Emmitsburg, MD, March 8-19, 1993.

    Research focuses on the problem of identifing costs related to fire station modification, due to introduction of women into the fire service.

3. Choatner, C. A., "Building Station as House Cuts Cost, Provides for Later Sale as Residence," Fire Engineering, Vol. 133(5), May 1980, pp. 32+.

Offers alternative approach for siting and building a fire station based on later intended use as residence.

4. Conner, Mitchell S. and John W. Rentz, "'No Surprises' Fire Station Design," Fire Chief, February 1992, pp. 60-62.

The benefits of cost modeling are described with examples from a station built in Novato, CA.

5. De Silva, Paul J., "From the Ground Up: Constructing or Reconstructing a Modern Firehouse" (part one of a series), Fire Engineering, October 1989, pp. 78-81.

Article presents an overview of the planning, architect selection, and bidding phases for new construction or remodeling.

6. De Silva, Paul J., "From the Ground Up: Constructing or Reconstructing a Modern Firehouse" (part three of a series), Fire Engineering, March, 1990, pp. 81-86.

Design considerations and special concerns such as explosion hazards and ventilation are addressed. Construction documents are discussed.

7. De Silva, Paul J., "From the Ground Up: Constructing or Reconstructing a Modern Firehouse" (part four of a series), Fire Engineering, May 1990, pp. 69-75

Funding options, project cash flow, and bidding are discussed.

8. De Silva, Paul J., "From the Ground Up: Constructing or Reconstructing a Modern Firehouse" (part five of a series), Fire Engineering, September 1990, pp. 73-75.

Project bidding and construction checklists are discussed.

9. DeWall, James S., "Components to Consider When Selecting An Architect," part of Executive Fire Officer Program Series, National Fire Academy, Emmitsburg, MD, March 1991.

Uses a department's experiences to illustrate recommendations for evaluating alternatives when hiring an architect.

10. Kamrath, Dale S., "Fire Station Construction - Efficient and Effective," part of Executive Fire Officer Program Series, National Fire Academy, Emmitsburg, MD, December 17, 1993.

Fire station construction is examined from a fiscal management perspective to determine what issues need to be addressed to provide for cost efficiency and effectiveness.

11. Moore, Bruce A., "Alternative Financing for Major Expenditures," part of Executive Fire Officer Program Series, National Fire Academy, Emmitsburg, MD, May 18, 1990.

Looked as ways of spreading payments over longer period of time for financing a new station construction project.

12. Murphy, John K., "Mitigation Funding Resolves Fire Protection Problems," Fire Chief, Vol. 34(8), August 1990, pp. 98.

Shows an example of cooperation between developers and fire departments to ensure adequate fire protection for new communities.

13. Quillin, Thomas C., "A More Efficient and Cost Effective Method of Providing Fuel to Units in Remotedly Located Fire Stations," part of Executive Fire Officer Program Series, National Fire Academy, Emmitsburg, MD, January 1994.

Examines the costs of fuel and methods for deliverying fuel to remote fire stations with the recommendation to install on-site fuel storage tanks as a secondary means of supplying fuel. The study recommends against obtaining fuel from private service stations.

14. Wallace, Lonzo, Fire Station Conservation Program," part of Executive Fire Officer Program Series, National Fire Academy, Emmitsburg, MD, December 1992.

Discusses one department's dilemma for reducing operating funds and personnel by downsizing, hiring overtime, and deactivating companies. Examines the advantages of a station conservation program in terms of utilities, supplies, and fuel.

15. Wenzel, James L., Increasing the Number of Fire Stations Without Hiring New Personnel," part of Executive Fire Officer Program Series, National Fire Academy, Emmitsburg, MD, June 29, 1993.

Examines the redistribution of personnel as a means for staffing a new fire station instead of hiring more personnel.

Station Design

1. ---,"Modular Construction Provides One-Engine Station in Four Days," Fire Chief, Vol. 23(10), October 1979, pp. 46-47.

Examines 70's based modular construction practices as applied to fire station construction.

2.   ---, "Fire, Medical Stations Cost Less in Pittsburgh," <u>ENR,</u> 18 Jan 1993, p. 21.

Briefly describes cost savings attributable to use of computer-based prototype station designs.

3.   ---, "Fire Stations," Design pamphlet for Dept. of Defense, November 1994.

Pamphlet includes standard designs for one and two company headquarters and satellite stations.

4.   ---, "What's New in Fire Stations," <u>Fire Chief</u>, Vol. 34(6), June 1990, pp. 33-35.

Shows new design ideas for apparatus bays, specialized areas, living quarters, storage areas, kitchen appliances, training areas, physical fitness areas based on interviews with 13 fire chiefs from paid, combination, and volunteer departments throughout the U.S.

5.   Amestoy, Lee, "Fire Station Construction - An Analysis of the Importance of Architectural Services," part of Fire Executive Development Series, National Fire Academy, Emmitsburg, MD, December 1991.

Provides the basis for how the choice of an architect can affect the fire station construction process.

6.   Aurnhammer, Thomas W., Mike Brodley, John Curtis, Harry Harris, Rodney Kelley, and Mike Wigdarson, "Cost-Efficient Fire Stations: Energy Saving Design," part of Fire Executive Development Series, National Fire Academy, Emmitsburg, MD, April 17-28, 1989.

Describes station design features which help cut construction and operating costs.

7.   Baker, George W., Peter J. Carver, Earl B. Gorrondona, Gregory J. Hanchar, Troy Glen Harris, James R. Patterson, and James V. Yates," Fire Stations.. .Today and Tomorrow, A Planning Tool," part of Fire Executive Development Series, National Fire Academy, Emmitsburg, MD, March 26-April 6, 1990.

Discusses need for flexibility in fire station design to accommodate future needs.

8.   Bedlington, John, "A Step-By-Step Guide to Designing Your Brigade's New Fire Station," &, Vol. 85(1046), pp. 17+.

Lists several considerations for designing fire stations in terms of site selection, layout, design features, and other concerns.

9. Bounds, A. and Rebecca Zurier, "The American Firehouse: An Architectural and Social History," 1st Edition, Abbeville Press, New York, 1982, pp. 282.

    Examines the role of fire station as part of the community and how station archictecture has evolved over the years in the United States.

10. Bryan, John L. and Raymond C. Picard, <u>Managing Fire Services,</u> International City Management Association, Washington, DC, 1979.

    Includes general practices for station location, design, and management as part of overall management strategy for the fire service.

11. Casey, James F., Donald L. Drumm, W. Thomas Schaardt, and Dick Silva, "The Complete Fire Station: Planning, Equipping, and Manning," Reprinted Report, Fire <u>Engineering,</u> October 1969, pp. 25.

    Focuses on station planning in terms of accommodating personnel and apparatus needs.

12. Coleman, Ronny, "A New Look into the Old Fire House," <u>Fire Chief,</u> November 1989, pp. 34-35.

    Briefly discusses selected design issues such as living quarters layout and maintenance.

13. Crosley, R. Stanley, "Fire Station Design: Justification of Space Requirements and Concepts for the Sidney, Ohio Fire Department," part of Executive Fire Officer Program Series, National Fire Academy, Emmitsburg, MD, August 1992.

    Indicates how one department prepared a plan for justifying their proposed station design to city management and the community.

14. Cross, Chris, "A Nuts and Bolts Renovation," <u>NFPA Journal,</u> Vol. 85(1) January-February 1991, pp. 94-99.

    Describes how a St. Louis department implemented a construction and renovation program to refurbish its stations for flexibility in meeting future needs.

15. De Silva, Paul, "Chapter 14 - Fire Station and Facility Design," and "Appendix 1 - Design Portfolio," <u>The Fire Chief's Handbook, Fifth Edition,</u> Joseph R. Bachtler and Thomas F. Brennan, editors, Fire EngineeringR Books and Videos, Pennwell Publishing Company, 1995, pp. 476ff.

Chapter 14 addresses all aspects of fire station and facility design, including site planning, layout, space considerations, budgets, and construction. Appendix 1 includes designs for eleven fire stations.

16.   De Silva, Paul J., "From the Ground Up: Constructing or Reconstructing a Modern Firehouse" (part two of a series), <u>Fire Engineering,</u> January, 1990, pp. 32-35.

Site selection and station layout are discussed for new or renovated firehouses. Special needs for apparatus bays and other facilities such as SCBA fill, Haz-mat, and training areas are addressed.

17.   De Silva, Paul J., "From the Ground Up: Constructing or Reconstructing a Modern Firehouse" (part three of a series), <u>Fire Engineering</u>, February, 1990, pp. 81-86.

Design considerations and special concerns such as explosion hazards and ventilation are addressed. Construction documents are discussed.

18.   Ely, Robert H., "Fire Station Planning, Design and Construction," published by International Association of Fire Chiefs Foundation, Washington, D. C., 1989.

Comprehensive guide for fire station guide complete with sample floorplans.

19.   Federal Aviation Administration, "Airport Rescue and Firefighting Station Building Design," FAA advisory circular, July 30, 1987.

Gives overview of design considerations for airport fire stations.

20.   Hanley, Paul, "Critical Facilities: Reducing Earthquake Hazards in the Central U.S. ", Report, University of Illinois at Urbana-Champaign, Urbana, IL, November 1992.

Cites construction building designs and construction methods for minimizing earthquake damage and ensuring survivability of critical facilities including fire stations.

21.   Marinucci, Richard A., "Near-Collapse at the Fire Station," <u>Fire Engineering</u>, Vol. 145, October 1992, pp. 67-68.

Fire retardant caused deterioration of wood trusses which resulted in complete breakage of the trusses, and near-collapse of the fire station's roof. The importance of routine inspections to uncover such problems is stressed.

22. Matthias, Fred T., "Fire Station Design," <u>Fire Chief</u>, Vol. 28(4), April, 1984, p. 74.

    Provides results from questionnaire to over 100 fire chiefs as to preferences for fire station design.

23. Navarre, Raymond J., PhD. , "Developing a Stress-reducing Fire Station," <u>Fire Chief</u>, February 1987, pp, 46-47.

    A general fire station design to reduce fire fighter stress is presented.

24. National Science Foundation, "Seismic Design for Police and Fire Stations." published by AIA Research Corp., Washington, D.C., 1978.

    Provides guidelines for building construction to allow stations to withstand earthquakes.

25. Peige, John, "Firehouse's Firehouses," <u>Firehouse</u>, Vol. 7(9), September 1982, pp. 59-63+.

    Pictorial review of attractive and traditional fire station designs.

26. Rentz, John W., "Building a New Fire Station," <u>Firehouse</u>, January 1989, p. 57.

    Successful incorporation of employees' input into the station design and special design features are described for a new California fire station.

27. Rice, Joe, "Truck Bays Angled in Texas Station," <u>Fire Engineering</u>, Vol. 131(6), June 1978, pp. 26-27.

    Provides alternative designs for locating apparatus bays for improved visibility and ease of backing.

28. Rondinelli, Stephen C., "Consolidated Fire Training Facility and Architectural Design Facility," part of Executive Fire Officer Program series, National Fire Academy, Emmitsburg, MD, April 1990.

    Describes the development of an architectural design program to identify the users' needs for a training facility for a suburban fire department.

29. Schaardt, W. Thomas and Roi Wooley, "Fire Station Design," American Insurance Association, National Board of Fire Underwriters, National Fire Protection Association, Warren, MI, 1973.

    Provides general overview of fire station design principles.

30. Sommerville, Thomas C., "The Vintage Firehouse: Gem's of America's Architectural Past," Firehouse, Vol. 15(7), July 1990, pp. 48+.

Examines traditional styles of fire stations dating back to the early 1800 's.

31. Trull, Elaine E., "SC Firehouse Blends With Victorian Architecture of Community," Firehouse, Vol. 13(6), pp. 110-111 +.

Shows the design of station for blending with community architecture.

32. U.S. Army Corps of Engineers, "Standard Fire Station Design," Report, Ellis Naeyaert Genheimer Associates, Troy, MI, not dated, 10 pps.

Provides drawings, design analysis, and narratives describing the recommended standard design for U. S. Army fire stations (for one and two company headquarters and satellite stations).

33. Wagner, Gary, "Old Time Station Design," Firehouse, Vol. 10 (5), May 1985, pp. 55-56.

Provides an example of how a new station can be designed using more traditional fire station design features.

34. Walker, Patrick L., Development of a Fire Station Design Guide," part of Executive Fire Officer Program Series, National Fire Academy, Emmitsburg, MD, October 1993.

Discusses how standards for the design and construction of fire stations have significantly changed in recent years in conjunction with changes in local building codes and technological developments.

35. Walsh, Michelle Byrne, "Evanston's New Station Blends Economy and Utility," Fire Chief, June 1990, pp. 36-38.

Briefly describes state-of-the-art attributes of a satellite fire station.

36. Walsh, Michelle B., "What's New in Fire Stations," Fire Chief, June 1990, pp. 33-35.

A panel of thirteen fire chiefs discuss a "wish list" of design features for the ideal station.

Station Safety

1. ---, "Fire Station Safety," Chapter 5 of Fire Department Occupational Safety, 2nd Edition, International Fire Service Training Association, 1991, pp. 99-117.

Station design. is discussed in reference to safety and efficiency. Lifting and fall/slip/trip hazards, poles and slides, housekeeping, electrical hazards, and other concerns are addressed.

2.   ---. "Safety Checklist for Firefighters," Appendix B of <u>Fire Department Occupational Safety, 2nd Edition</u>, International Fire Service Training Association, 1991, pp. 339-342.

Appendix includes checklists summarizing safety considerations for the interior and exterior of the fire station.

3.   Bard, George R., "Is Your Fire House Safe?" <u>Health & Safety for Fire and Emergency Service Personnel</u>, Fire Department Safety Officers Association, June 1995, pp. 1,4.

Briefly describes basic components of an installed alarm system for a fire station and includes a check list for a self-inspection program.

4.   Mesagna, Emmanuel and John Baroni, "Fire Department Facilities and Fire Training Facilities," Section 9/Chapter 10 of the <u>Fire Protection Handbook.TM -Seventeenth Edition</u>, Arthur E. Cote and Jim L. Linville, Editors, National Fire Protection Association, Quincy, Massachusetts, 1991, pp. 9-100 to 9-108.

Site selection, concept planning, and space planning analysis are discussed.

5.   Parker, T. J., "Rocky Flats Scores a Hit with PITS," <u>American Fire Journal</u>, July 1995, pp. 22-24.

Article describes the implementation of "Process Improvement Teams," a team approach to evaluating safety aspects of the activities and jobs that comprise fire fighters' jobs.

6.   Perry, Dale, "Maintenance Policies: Penny Wise and Pound Foolish?," <u>Health & Safety</u>, Vol. 6(6), June 1995, pp. 8-10.

Cutting corners on repairing fire apparatus can be costly in the long run. This article cites evidence of what can go wrong when unskilled personnel try to repair a fire pumper's gas tank.

7.   Sachs, Gordon M., "Safety Begins in the Fire Station," <u>Fire Command</u>, July, 1988, pp. 15-17+.

Various hazards in the fire station, including diesel emissions, and aspects of a facility protection program are discussed.

8.   Simmons, Tim, "Dream Station," Phoenix Fireworks, Vol. 10, August 1987, p. 1,3.

Layout is described for a fire station incorporating a Haz-Mat team, heliport, and training facility.

9.   Thorpe, Tom, "How Safe Is Your Fire Station," Fire Engineering, Vol. 147, April 1994, pp. 79-82 +.

Article addresses fire station safety hazards and issues, including the hose tower, fall hazards, electrical hazards, slide poles, and pressurized vessels and cylinders.

Fires/Accidents at Fire Stations

1.   ---,"There's Nothing Left! Explosion Levels Parkers Prairie. Station," Minnesota Fire Chief, Vol. 31(5), May-June 1995, p.7

Describes circumstances leading to and results of explosion at Minnesota fire station.

2.   Berthinier , James, "Retrofitting Existing Fire Stations with Automatic Sprinkler Systems," part of Executive Fire Officer Program Series, National Fire Academy, Emmitsburg, MD, January 1991.

Raises the concern that use of automatic sprinkler systems at fire stations should be a priority particularly to heighten community use of sprinkler systems.

3.   Bradish, Jay K., "Worst Nightmare' Becomes a Reality," Firehouse, June 1995, pp. 88, 90.

Response to and damages caused by a fire at a volunteer fire station are described.

4.   Ellyson, Richard V., "Description of Need and Adaptation of an Automatic Fire Extinguisher Unit for Residential Kitchen Range Top Protection in a Fire Station," part of Executive Fire Officer Program series, National Fire Academy, Emmitsburg, MD, May 1992.

Gives examples of fires caused as stations by ranges and stoves left on during responses and how modified electrical safety systems can reduce these hazards.

5.   Nessler, Dennis, "You Never Think It's Going to Happen to You," Firehouse, June 1995, pp. 84-86.

Response to and damages caused by a fire at an unmanned firehouse are described.

6.   Wallace, Stephen, "It Can Happen to Anyone," <u>Fire Chief</u>, Vol. 36(7), July 1992, pp. 30.

A preliminary investigation of a fire station fire revealed that hot grease caught fire thrusting it into the exhaust van. The fan melted and ignited and the fire then worked its way up the wall. Fire personnel existing the ready room to respond to an alarm forgot that the burner under a greasy pan was still on.

7.   Wood, Thomas R., "A Study of Structure Fires and Fire Sprinkler Systems in the U. S. Fire Stations," part of the Executive Fire Officer Program Series, National fire Academy, Emmitsburg, MD, May 1992.

This report provides results of research aimed at answering the following questions: (1) How many structure fires occur in U.S. fire stations?, (2) what are the more common causes of these fires?,(3) how many stations have fire sprinkler protection?, and (4) would properly installed fire sprinkler systems reduce fire damage losses in U.S. fire station fires ?

8.   Wood, Thomas R., "Survey: Fire Stations No Stranger to BLazes," <u>IAFC On Scene</u>, Vol. 6 (17), September 15, 1992, pp. 6-7.

Provides survey results for number of fires, their causes, and number of sprinkler systems in U.S. fire stations.

Special Issues

1.   ---, "Fire Station Saves Money and Energy," <u>International Fire Chief</u>, Vol. 47(6), June 1981, pp. 19.

Provides ideas for reducing fire station construction budget by looking at energy efficient designs.

2.   Amberge, George, "Low-Cost Automatic Lighting for Fire Stations," <u>Fire Engineering</u>, September, 1991, p. 60-61.

Article describes an affordable signal-activated lighting system suitable for fire station use. The system consists of plug-in sending and receiving modules that activate lamps for a preset period when an alarm is sounded.

3.   Blackistone, Stephen D., "Sexual Harrassment in the Fire House," <u>Firehouse</u>, Vol. 17(4), April 1992, pp. 90-92.

Reports on growing concerns about sexual harrasement at fire stations from increased participation of females in the fire service.

4.   Blackistone, Stephen D., "Sexual Harrassment Versus Privacy: Feeling Comfortable in the Firehouse," Firehouse, Vol. 20(5), May 1995, pp. 103-104.

Addresses issues such as employee rights, gender differences in fire stations and the need for departments to accommodate these concerns in station design.

5.   Carpentino, Frank M., Charles D. Harding, David W. House, Kurt P. Larson, and V. Dominic, "Determining Whether the Vigilant Hose Company Fire Station Complies with Current Fire Station Design Criteria as it Relates to the American With Disabilities Act of 1990 and NFPA 1500, 1992 Edition," part of Executive Fire Officer Program Series, National Fire Academy, Emmitsburg, MD, May 3-14, 1993.

This study shows how an existing station can be inspected for compliance with ADA and NFPA 1500.

6.   Chapman, Suzanne, "$1.5 Million AFB Fire Station Has Solar Energy Domestic Hot Water," Fire Engineering, Vol. 133(11), November 1980, p. 26.

Shows how solar-based water heating system reduces energy costs for modern fire station.

7.   Kluck, Timothy J., Allyn Lee, Leonard L. Lowen, David Munger, Peter S. Polarek, Michael A. Springer, " part of Executive Fire Officer Program Series, National Fire Academy, Emmitsburg, MD, May 23-June 3, 1994.

Provides survey results for more than 86 departments on how fire departments deal with gender-based issues including sleeping quarters.

8.   Nailen, R. L., "Sun Helps Heat, Cool Dallas Station," Fire Engineering, Vol. 131(3), March 1978, pp. 26-27.

Provides example of how solar energy effectively reduces energy costs at one station.

9.   Rhodus, Dan, "Americans with Disabilities Ace Title II: Lenexa Fire Department Facility Compliance," part of Executive Fire Officer Program Series, National Fire Academy, Emmitsburg, MD, May 1992.

Establishes need for fire departments to address ADA Title II compliance requirements and examines typical examples of how stations fail to comply.

10.   Schneid, Thomas D., "Is Your Fire Station Safe From Workplace Violence?" Fire Engineering, August 1995, pp. 137

Presents an overview of workplace violence and prevention programs.

11.   Silvestri, Richard C. , "Human Needs in Constructing Fire Stations," part of Executive Fire Officer Program Series, National Fire Academy, Emmitsburg, MD, November 1992.

Discusses need for ergonomic approach to fire station design.

12.   Tubbs, R. L., "Occupational Noise Exposure and Hearing Loss in Fire Fighters Assigned to Airport Fire Stations," American Industrial Hygiene Association Journal, Vol. 52 (9), September 1991, pp. 372-8.

Applicable topic: firefighter hearing loss due to occupational exposure

13.   Willing, Linda, "Bedrooms & Bathrooms I: The Hidden Message," WFS Quarterly, Winter, 1988-1989, p. 1-2.

Female fire fighter issues about station facilities are discussed.

Infection Control

1.   Bizjak, Gloria, "Contamination in Fire Service Workplaces," Maryland Fire and Rescue Bulletin, December 1989, pp. 7-8.

Disease transmission routes are discussed. Laundry and food handling concerns are outlined.

2.   Landrigan, Philip J., MD, and Dean B, Baker, MD, "The Recognition and Control of Occupational Disease," Journal of the American Medical Association, Vol. 266 (5), 7 August 1991, pp. 676-680.

Describes practices for preventing workplace exposures

3.   Mueller, Kim, "Prevention and HIV Transmission in the Workplace: Category II (Emergency Service) Workers," American Industrial Hygiene Journal, Vol. 52 (2), February 1991, pp. 104-107.

Provides guidelines for departmental HIV policies

4.   Strickland, David, "Critics Assail OSHA Regulations on Protection against HBV and HIV," Medical World News. 9 October 1989, pp. 14-15.

   Offers comments on laundry and training for reducing bloodborne pathogen exposures and for complying with 29 CFR 1910.1030.

5.   West, Katherine H., "OSHA's Final Work on Bloodborne Pathogens," Fire Engineering, April 1992, pp. 82-84.

   Provides suggestions for fire departments for implementing OSHA regulations at the station level.

6.   Williams, Decker and Dean Pedrotti, "Blueprint for a Disease-Free Station," Firehouse, February 1988, pp. 70, 73.

   Design features and procedures are described for reducing exposure in the fire station.

Diesel Emission/Indoor Air Quality

1.   Adams, Roger W., "Health Concerns of Passive Smoke in Fire Stations," part of Executive Fire Officer Program Series, National Fire Academy, Emtnitsburg, MD, August 1993.

   The research for this report determines if the workplace should be free of passive smoke, identifies what policies are currently in place within Oregon fire departments, and what strategies can be utilized to implement a smoke-free workplace policy.

2.   Curtis, Jeffrey M., John C. Huff, Glenn S. Mutch, C. Dan Rhodus II, William J. Seng, and Danny R. Shockley, "Reducing and Removing Diesel Emission in Fire Station Living Areas," submitted to the National Fire Academy as part of the Executive Fire Officer Program, December, 1991.

   Outlines problems and alternative solutions for control of diesel exhaust at the fire station.

3.   Froines, John R., William C. Hinds, Richard M. Duffy, Edward J. Lafuente, and Wenthen V. Liu, "Exposure of Firefighters to Diesel Emissions in Fire Stations," American Industrial Hygiene Association Journal, Vol. 48, March 1987, pp. 202-207.

   Findings from a study of firefighter exposure indicated increased exposure toparticulates from diesel engine emissions. Personal sampling techniques were used in the evaluations conducted in three cities.

4.   Girod, Gary, "Benzene in Diesel Exhaust Increase," Speaking of Fire, Fall 1989, pp. 4-5.

     Reports findings for air analysis at several fire stations.

5.   Gulick, John G., Exhaust Emissions from Fire Apparatus - Potentially Harmful to Fire Personnel," part of Executive Fire Officer Program Series, National Fire Academy, Emmitsburg, MD, January 1994.

     Explains the long term health effects associated with the carcinogenicity of diesel exhaust fumes and addresses the effectiveness of current solutions for control diesel emissions at fire stations.

6.   Kinsey , William C., Jr., "Firefighters and Diesel Emissions: A Current Study of Workplace Exposure," submitted to the National Fire Academy as Part of the Executive Fire Officer Program, August, 1991.

     Reports air analysis findings for stations in one department and considerations for selecting methods for diesel exhaust control.

7.   New Jersey State Department of Health, "Diesel Exhaust in Fire Stations," Information Bulletin, September, 1986.

     Announces health effects of diesel exhaust and puts fire stations on notice of possible carcinogen status.

8.   Parks, Kenneth G., "Reducing Exposure of Personnel to Diesel Exhaust Emission at the Baltimore County Fire Department, Catonsville Station #4," submitted to the National Fire Academy as part of the Executive Fire Office Program, August, 1989.

     Discusses implementation of source capture system for removal of diesel exhaust emissions.

9.   Peters, William C., "Diesel Soot: An Exhausting Problem," Fire Engineering, March 1992, pp. 47-51.

     Describes hazards of diesel exhaust emissions and recommends different methods of source capture for its control.

10. Pfeiffer, Edward A., "Don't Let the Smoke Get in Your Eyes," Fire Chief, April 1992, pp. 126-7.

     Provides an overview of diesel emission problems at the fire station,

11. Porter, Coy D., "Are We Killing Ourselves? - A Comparative Study of the Effects of Station House Smoke as it is Relates to the Health of Fire Fighters," part of Executive Fire Officer Program Series, National Fire Academy, Emmitsburg, MD, July 1993,

     Consolidates information on the toxic effects of diesel emission exposures and second-hand (cigarette) smoke typically found in fire stations around the country.

12. Roseboro, Christopher, "Reducing Fire Fighter Exposure to Diesel Emissions," part of Executive Fire Officer Program Series, National Fire Academy, Emmitsburg, MD, December 1992.

     Research in project demonstrated and described the three general ways for reducing diesel exhause exposure including engineering controls, general ventilation, and source capture. The study concluded that the better and more efficient the solution, the more complicated and longer term the fix.

13. Sachs, Gordon M., "Safety Begins in the Fire Station," <u>Fire Command,</u> Vol. 55, July 1988, pp. 15-17+.

     Addresses application of NFPA standards to safety in the fire station, labels diesel exhaust as major station pollutant.

14. Stull, Jeffrey 0., "Controlling Diesel Exhaust Emissions at the Fire Station," Fire <u>Engineering</u>, Vol. 147, October 1994, pp. 18+.

     Discusses diesel engine emission hazards and exhaust control strategies with recommendation of source capture as preferred method of control.

15. Van Atta, Keith A., "Strategies for Reducing Diesel Exhaust Exposure of Fire Fighters," part of Executive Fire Officer Program Series, National Fire Academy, Emmitsburg, MD, August 1992.

     Describes a study intended to evaluate the state of fire service in Oregon by determining which fire departments have already adopted strategies, what the strategies were, and how they decided on the strategies for reducing diesel exhaust emissions.

16. Weixeldorfer, Edward R., "The Monitoring of Diesel Exhaust Emissions in Three Selected Kansas City, Missouri Fire Stations," part of Executive Fire Officer Program Series, National Fire Academy, Emmitsburg, MD, August 1992.

     Air monitoring data were collected from three Kansas City fire stations to determine the amounts of carbon monoxide and respirable particulate dust during three days of station operations, The results showed higher than normal levels for these air contaminants.

APPENDIX F

EXAMPLE STATION DESIGN
SAFETY AND HEALTH CHECK-OFF LIST

APPENDIX F. EXAMPLE STATION DESIGN SAFETY AND HEALTH CHECK-OFF LIST

No.	Area/Inspection Item	Inspection Results			Comments and Recommended Changes
		Yes	No	N/A	
Site Layout					
1	Phase I environmental site assessment conducted				
2	Station with "one way" circulation drive, if possible				
3	Adjacent street traffic areas equipped with traffic control lights and vehicle stop bars				
4	Turns less than 90 degrees to enter public street				
General Site Construction					
1	Fire resistance ratings for structure meet local building codes				
2	Active soil depressurization system installed or similar technique used for areas where radon hazard exists				
3	Structural design consistent with earthquake resistance (seismic performance) as appropriate for area				
All Areas					
1	Electrical outlets grounded and connected to electrical panel with appropriate sized circuit breaker.				
2	Circuit breakers clearly identified				

Examine each space of the station design or existing station and mark compliance as yes, no, or non-applicable. In comments section, indicate type of non compliance and recommended action/design change to correct non-compliance.

F-2

APPENDIX F. EXAMPLE STATION DESIGN SAFETY AND HEALTH CHECK-OFF LIST (Continued)

No.	Area/Inspection Item	Inspection Results			Comments and Recommended Changes
		Yes	No	N/A	
3	Electrical panel area clear and free of storage				
4	Electrical panel provided with cover				
5	All electrical receptacles and junction boxes covered with appropriate plates				
6	Ground fault interrupt circuits installed in wet areas (including bathrooms, apparatus bays, boiler rooms, rooms, outside areas, and kitchens)				
7	Sufficient receptacles in each space				
8	Electrical system installed and maintained by a competent electrician				
9	All electrical appliance grounded and kept clean				
10	All electrical motors adequately ventilated and cleaned regularly				
11	Cage or other protection provided for lights within 7 feet of floor				
12	Smoke detectors installed into main electrical system in each area				
13	Carbon monoxide sensors installed in each living space				

Examine each space of the station design or existing station and mark compliance as yes, no, or non-applicable. In comments section, indicate type of non compliance and recommended action/design change to correct non-compliance.

F-3

APPENDIX F. EXAMPLE STATION DESIGN SAFETY AND HEALTH CHECK-OFF LIST (Continued)

No.	Area/Inspection Item	Inspection Results			Comments and Recommended Changes
		Yes	No	N/A	
14	Separate, non-confusing fire alarm system installed				
15	Portable fire extinguishers provided in major spaces				
16	Automatic sprinkler system installed				
17	Number of exits consistent with size of structure and expected maximum occupancy				
18	Exits located remotely from each other				
19	Exits conspicuously marked with illuminated "Exit" sign				
20	Exit doors a minimum of 28 inches wide and 7 feet 6 inches high				
21	Clear, full length glass doors and windows properly marked				
22	Stairway and exit doors kept closed at all times				
23	Alarm/security system installed				
24	Alarm/security system contacts and sensors well hidden				
25	Control system for a arm/security system in locked room or closet away from high traffic areas				

Examine each space of the station design or existing station and mark compliance as yes, no, or non-applicable. In comments section, indicate type of non compliance and recommended action/design change to correct non-compliance.

APPENDIX F. EXAMPLE STATION DESIGN SAFETY AND HEALTH CHECK-OFF LIST (Continued)

No.	Area/Inspection Item	Inspection Results			Comments and Recommended Changes
		Yes	No	N/A	
26	Non-flammable cleaning agents used throughout the building				
27	Ready disposal of combustible wastes provided				
28	Trash and rubbish stored in metal containers				
Station Grounds					
1	Utility poles space 10 to 15 feet from vehicle maneuvering areas				
2	Ground mounted transformers protected from impact when 10 feet from driveways or parking lots (a 1 hour-rated fire wall is recommended)				
3	Ground fault interrupt circuits installed				
4	Driveways (circulation pathways, training areas, and vehicle aprons) levels, well drained and use non-slip surfaces rated for weight of apparatus				
5	Non-slip texture applied to concrete walkways and other surfaces				
6	Painting minimized for asphalt surfaces				

Examine each space of the station design or existing station and mark compliance as yes, no, or non-applicable. In comments section, indicate type of non compliance and recommended action/design change to correct non-compliance.

APPENDIX F. EXAMPLE STATION DESIGN SAFETY AND HEALTH CHECK-OFF LIST (Continued)

No.	Area/Inspection Item	Inspection Results			Comments and Recommended Changes
		Yes	No	N/A	
7	Checked metal plate or other non-slip surface covering for underground vaults, test pits, or transition surfaces				
8	Sufficient parking provided which is secure and well lighted				
9	Exterior lighting adequate in all areas				
10	Non-slip surfaces applied for drill towers and other heavy traffic areas				
11	Run-off basins with oily-water separators provided for apparatus washing areas				
12	Above storage tanks chosen before underground tanks				
13	Secondary containment for above ground storage tank equals volume of primary storage tank				
14	System in place to prevent back flow and cross contamination				
15	Noise or sound control breaks (buffer zone) placed on sides of station property				
16	Exterior fire escapes in good condition				
17	Pedestrian warning signs posted next to station				

Examine each space of the station design or existing station and mark compliance as yes, no, or non-applicable. In comments section, indicate type of non compliance and recommended action/design change to correct non-compliance.

F-6

APPENDIX F. EXAMPLE STATION DESIGN SAFETY AND HEALTH CHECK-OFF LIST (Continued)

No.	Area/Inspection Item	Inspection Results			Comments and Recommended Changes
		Yes	No	N/A	
18	Storage and trash at least 25 feet away from building				
19	Heating provided under sidewalks and other areas prone to freezing				
20	Door buzzer or bell for main entry door				
21	Vandalism and violence proof windows installed				
Apparatus Bays					
1	Ground fault interrupt circuits installed				
2	Provide at least 3 feet of clearance around each vehicle in the apparatus bay				
3	Provide non-slip surface on apparatus bay floor particularly in areas where personnel mount or dismount apparatus				
4	Self-closing doors installed which have pressure switches to prevent door accidents				
5	Painted lines on apparatus bay floor for guiding vehicle				
6	Ventilation system or source capture system provide for control of diesel exhaust gases				

Examine each space of the station design or existing station and mark compliance as yes, no or non-applicable. In comments section, indicate type of non compliance and recommended action/design change to correct non-compliance.

APPENDIX F. EXAMPLE STATION DESIGN SAFETY AND HEALTH CHECK-OFF LIST (Continued)

No.	Area/Inspection Item	Inspection Results			Comments and Recommended Changes
		Yes	No	N/A	
General Station Interior					
1	Access from quarters to apparatus bays in straight line (hallways and crossing areas avoided)				
2	Access ways to the apparatus bay at front or rear of apparatus bay				
3	Separate rooms designated for smoking or smoking prohibited except outside station in area away from fuel or other flammable hazards				
4	Ventilation in interior living spaces at a positive pressure relative to apparatus bay *(optional)*				
5	Halls and corridors a minimum of 5 feet wide				
6	Building entrances compliant with ADA requirements				
Heating and Air Conditioning Equipment					
1	Heating equipment sized appropriate for structure				
2	Heating and air conditioning equipment inspected by HVAC service person each year				
3	Heating equipment properly insulated from combustible materials				

Examine each space of the station design or existing station and mark compliance as yes, no, or non-applicable. In comments section, indicate type of non compliance and recommended action/design change to correct non-compliance.

APPENDIX F. EXAMPLE STATION DESIGN SAFETY AND HEALTH CHECK-OFF LIST (Continued)

No.	Area/Inspection Item	Inspection Results			Comments and Recommended Changes
		Yes	No	N/A	
4	Heating and air conditioning equipment spaces restricted areas				
5	Heating and air conditioning equipment spaces free of storage				
Quarters					
1	If no smoking is permitted, posted signs: "No Smoking"				
2	Has proper lighting				
3	Separated from noisy areas of the station				
4	Individual lockers provided for all personnel with lock				
Watch/Dispatch Area					
1	Controls for traffic lights, alarms				
2	Switches for fuel pumps				
3	Slip resistance floor surface				
4	Has proper lighting				

Examine each space of the station design or existing station and mark compliance as yes, no, or non-applicable. In comments section, indicate type of non compliance and recommended action/design change to correct non-compliance.

APPENDIX F. EXAMPLE STATION DESIGN SAFETY AND HEALTH CHECK-OFF LIST (Continued)

No.	Area/Inspection Item	Inspection Results			Comments and Recommended Changes
		Yes	No	N/A	
Offices					
1	Vertical files with interlock allowing only one drawer open at a time				
2	Has proper lighting				
3	Separated from noisy areas of the station				
Classrooms					
1	Separated from noisy areas of the station				
2	Has proper lighting				
Kitchen					
1	Ground fault interrupt circuits installed				
2	Alarm-activated service disconnect of all fixed cooking devices installed (optional)				
3	Kitchen appliance protected with automatic fire extinguishing system (optional)				
4	Hood and duct system installed properly and regularly cleaned				

Examine each space of the station design or existing station and mark compliance as yes, no, or non-applicable. In comments section, indicate type of non compliance and recommended action/design change to correct non-compliance.

No.	Area/Inspection Item	Inspection Results			Comments and Recommended Changes
		Yes	No	N/A	
5	Grease filters UL listed and for grease extraction and properly installed.				
6	Ranges, dishwashers, and refrigerators are of commercial grade				
7	Double sinks with heavy duty garbage disposals provided				
8	Seperate storage cabinets for food storage				
9	Kitchen layout with center island counter (optional)				
Bathrooms					
1	Seperate gender-specific facilities offered				
2	Seperate toilet facilities for public (optional)				
3	Ground fault interrupt circuits installed				
4	Non-slip strips or rubber mats applied on shower floors				
5	Non-porous, easily cleaned surfaces for shower stalls and walls				
6	System in place to prevent back flow and cross contamination				

Examine each space of the station design or existing station and mark compliance as yes, no, or non-applicable. In comments section, indicate type of non compliance and recommended action/design change to correct non-compliance.

APPENDIX F. EXAMPLE STATION DESIGN SAFETY AND HEALTH CHECK-OFF LIST (Continued)

No.	Area/Inspection Item	Inspection Results			Comments and Recommended Changes
		Yes	No	N/A	
Support Areas (e.g. machine shops)					
1	All electrical powered tools or equipment insulated				
2	All electrical outlets have ground wire				
3	Block heaters feed from overhead electrical location				
Disinfection Areas					
1	Separate space provided for disinfection				
2	Has ventilation to outside				
3	Has proper lighting				
4	Drains connected to a sanitary sewer system				
5	Disinfection area has minimum of 2 sinks with hot and cold water faucets and sprayer attachment				
6	Sink faucets designed for hand-less operation				
7	Sink and adjacent surfaces non-porous material with continuous molded counter top and splash panels				
8	Equipped with rack shelving of non-porous material for drip-drying equipment				

Examine each space of the station design or existing station and mark compliance as yes, no, or non-applicable. In comments section, indicate type of non compliance and recommended action/design change to correct non-compliance.

F-12

APPENDIX F. EXAMPLE STATION DESIGN SAFETY AND HEALTH CHECK-OFF LIST (Continued)

No.	Area/Inspection Item	Inspection Results Yes	No	N/A	Comments and Recommended Changes
9	Drainage from rack goes into sink or sanitary sewer system				
10	Front loading washing machines used for cleaning of protective clothing (optional)				
11	Waste water stored in double wall tank				
Cleaning Areas					
1	Seperate space provided for cleaning of PPE				
2	Has ventilation to outside				
3	Has proper lighting				
4	Drains connected to a sanitry sewage system				
5	Area physically seperate from food preparation, kitchen sleeping, living, disinfection areas				
6	Front loading washing machines used for cleaning of protective clothing (optional)				
Exercise Areas					
1	Adequate floor provided for exercise equipment				

Examine each space of the station design or existing station and mark compliance as yes, no, or non-applicable. In comments section, indicate type of non compliance and recommended action/design change to correct non-compliance.

APPENDIX F. EXAMPLE STATION DESIGN SAFETY AND HEALTH CHECK-OFF LIST (Continued)

No.	Area/Inspection Item	Inspection Results			Comments and Recommended Changes
		Yes	**N o**	**N/A**	
2	Only department approved exercise equipment in place				
3	Sign posted: "Buddy System Required for Weight Lifting"				
4	Sign posted: "Use Proper Lifting Techniques"				
Stairways					
1	Guardrails placed on open side so stairways where there are more than 4 rises				
2	Stairway rail height between 30 and 34 inches (measured from forward edge of tread to upper surface of top rail)				
3	Handrails no more than 2 inches in diameter				
4	Handrails height between 34 and 38 inches				
5	Handrails continuous and extend beyond the last step				
6	Guardrails with top rails at 42 inches high				
7	Guardrails with mid-rails at 21 inches high				
8	Non-slip finishes or tread on stair treads and tread nosing				
9	Color on stair contrasts with rest of tread				

Examine each space of the station design or existing station and mark compliance as yes, no, or non-applicable. In comments section, indicate type of non compliance and recommended action/design change to correct non-compliance.

APPENDIX F. EXAMPLE STATION DESIGN SAFETY AND HEALTH CHECK-OFF LIST (Continued)

No.	Area/Inspection Item	Inspection Results			Comments and Recommended Changes
		Yes	No	N/A	
Sliding Poles					
1	Guard provided around pole hole to prevent falling through (automatic doors may also be provided)				
2	3 foot diameter cushioned mate at bottom of slide pole				
3	Sign posted that pole is used by one person at a time				
Hose Towers and Other Elevated Platforms					
1	Guardrail capable of withstanding a force of 250 lbs applied in any dircetion at any point on the top rail				
2	Toe guards equipped on each platform				
3	Offset platforms and cage guards equipped on platforms when acces ladders extend beyond 20 feet (permanent fixed ladders on outside of drill towers and buildings aret exempt)				
4	Rung clearance of at least 7 inches behind hosae tower ladder				
5	Step-across distances not greater than 12 inches				
6	Side rails greater than 16 inches wide				

Examine each space of the station design or existing station and mark compliance as yes, no, or non-applicable. In comments section, indicate type of non compliance and recommended action/design change to correct non-compliance.

F-15

APPENDIX F. EXAMPLE STATION DESIGN SAFETY AND HEALTH CHECK-OFF LIST (Continued)

No.	Area/Inspection Item	Inspection Results			Comments and Recommended Changes
		Yes	No	N/A	
7	Ladder rung spacings less than 12 inches				
8	Ladder cases between 7 and 8 feet above landing				
9	Ladder pitch between 75 and 90 degrees				
Storage Areas					
1	Located on same level as living and working spaces				
2	Personal protective equipment stored away from apparatus bay in well-ventilated room				
3	Silhouette board provided for all tools				
4	Ventilation installed for spaces involving hazardous materials				
5	Explosive gas monitors or sensors installed in spaces involving hazardous and potentially explosive materials				
6	Pressurized cylinders secured to prevent overturning				
7	Verticle files with interlock allowing only one drawer open at a time				
8	½ inch lip on open storage shelving or high rise storage systems				

Examine each space of the station design or existing station and mark compliance as yes, no, or non-applicable. In comments section, indicate type of non compliance and recommended action/design change to correct non-compliance.

F-16

No.	Area/Inspection Items	Inspection Results			Comments and Recommended Changes
		Yes	No	N/A	
9	½ to 1 inch lip of wall mounted items				
10	42 inch high walls or rails on all mezzanine storage areas				
11	Access to mezzanine areas by 4 foot wide sliding or rolling gate				
12	Walled off loft spaces with access gate or door that has three restraining cables, chains, or bars				
13	Toe boards on all elevated storage areas				
14	Chemicals in original, labelled containers and segregated by hazard				
15	Material Safety Data Sheets (MSDS) provided on each chemical at station and kept in central location				
16	Location of Material Safety Data Sheets posted				
17	Emergency medical supplies and equipment stored in dedicated space				
18	Secure spaces provided for storage of sensitive medical supplies (e.g., drugs)				
19	Sufficient storage space for all supplies				

Examine each space of the station design or existing station and mark compliance as yes, no, or non-applicable. In comments section, indicate type of non compliance and recommended action/design change to correct non-compliance.

APPENDIX F. EXAMPLE STATION DESIGN SAFETY AND HEALTH CHECK-OFF LIST (Continued)

No.	Area/Inspection Item	Inspection Results			Comments and Recommended Changes
		Yes	No	N/A	
Refueling Areas					
1	Ventilation installed (for interior spaces)				
2	Explosive gas monitors or sensors installed				
3	Refueling pumps installed in accordance with local requirements				
4	"No Smoking - Stop Your Motor" signs posted				
5	Refueling pump shut-off switch placed a minimum of 50-60 linear feet away with sign, "Fuel Pump Shut-Off"				
Pressurized cylinders					
1	Pressure vessels (e.g., boilers, large air-tanks) inspected and certified				
2	Pressurized cylinders hydrostatically tests at regular intervals				
3	Pressurized cylinders secured to prevent overturning				

Examine each space of the station design or existing station and mark compliance as yes, no, or non-applicable. In comments section, indicate type of non compliance and recommended action/design change to correct non-compliance.

APPENDIX F. EXAMPLE STATION DESIGN SAFETY AND HEALTH CHECK-OFF LIST (Continued)

No.	Area/Inspection Item	Inspection Results			Comments and Recommended Changes
		Yes	No	N/A	
Marine Areas					
1	Dock or pier constructed of non-slip materials				
2	Non-slip finish applied to concrete areas along water				
3	Tie-off points, hose bibs, and electrical outlets located away from walkways				
4	Gangways designed for multiple angles of use with non-slip surfaces				

Examine each space of the station design or existing station and mark compliance as yes, no, or non-applicable. In comments section, indicate type of non compliance and recommended action/design change to correct non-compliance.

☆ U. S. GOVERNMENT PRINTING OFFICE: 1997-519-343/90013